The Role of the Measurement and Verification Professional

Judgment and Decision-making in the Application of M&V

The Role of the Measurement and Verification Professional

Judgment and Decision-making in the Application of M&V

Steve Kromer

SKEE, United States

River Publishers

Routledge
Taylor & Francis Group
NEW YORK AND LONDON

Published 2024 by River Publishers
River Publishers
Alsbjergvej 10, 9260 Gistrup, Denmark
www.riverpublishers.com

Distributed exclusively by Routledge
605 Third Avenue, New York, NY 10017, USA
4 Park Square, Milton Park, Abingdon, Oxon OX14 4RN

The Role of the Measurement and Verification Professional / by Steve Kromer.

Routledge is an imprint of the Taylor & Francis Group, an informa business

ISBN 978-87-7004-105-8 (hardback)
ISBN 978-10-4009-057-2 (online)
ISBN 978-10-0349-819-3 (master ebook)

Contents

Foreword

It is with great excitement that we announce the establishment of AEE Press with the launch of this inaugural publication "The Role of the Measurement and Verification Professional: Judgment and Decision-Making in the Application of M&V". This book represents the first of a collection that aims to educate and advance energy efficiency professionals by bringing together the knowledge and insights of experts dedicated to advancing the implementation and measurement of energy efficiency projects and programs. With the debut of this current and forward-thinking book, we lay the foundation for a series that promises to be a beacon of knowledge, fostering innovation and excellence in energy efficiency.

This book stands as a guiding tool for individuals on the path of becoming a Certified Measurement and Verification Professional® (CMVP ®) or for those who seek to enhance their understanding of the important world of measurement and verification. In a rapidly evolving era where energy efficiency, sustainability and decarbonization have become integral to global agendas, the role of measurement and verification professionals has never been more vital. As society strives to reduce its carbon footprint and conserve resources, the need for reliable measurement and verification becomes paramount. CMVPs act as the authority of reliability, quantifying energy conservation and ensuring energy savings are substantiated.

Becoming a Certified Measurement and Verification Professional is not merely a professional pursuit; it's a commitment to the performance of proper measurement verification. This process is characterized by an exploration of concepts spanning energy measurement, verification protocols, data analysis, and the complexities of various industries. Throughout this book, you will learn the methodologies and insights that empower CMVPs to navigate complex projects, champion innovation, and foster a culture of sustainability.

Whether you are an aspiring CMVP, a seasoned measurement and verification professional seeking to refine your skills, or an individual curious about the mechanisms that underpin the measurement and verification landscape, this book will serve as an invaluable resource. This book will prepare you to grasp the principles, methodologies, and practical applications that define the CMVP domain and acquire the tools to navigate the intricate domain of energy efficiency projects with confidence and proficiency.

Remember that your pursuit of the CMVP certification extends beyond personal achievement. It is a tribute to the collective endeavor to measure and verify energy savings. May this book navigate your path, enhance your knowledge, and inspire your dedication to becoming a Certified Measurement & Verification Professional.

Together, let us embark on a journey that not only elevates careers but also elevates the world toward a brighter, more sustainable clean energy future.

Bill Kent
Executive Director at Association of Energy Engineers

Preface

"Energy drives economies and sustains societies."[1] This simple statement has vast implications. Most obviously, it indicates just how dependent our world is on the utilization of energy.

Energy management is a practical means of ensuring that the costs and externalities of energy use are understood and mitigated. Common-sense energy management involves applying the fundamental laws of physics to energy systems with a goal of optimizing the value (services) produced per unit of energy consumed.

Energy management and related projects require an investment. And investments are made with the expectation of a return. Stakeholders must establish the value of energy management. In the context of both individual contracts and large-scale programs, stakeholders seek a quantitative answer to the question: "Did the energy management investment pay off?"

Stakeholders can measure the output of energy *supply* sources easily – by connecting a meter to a solar panel array or wind turbine, for example. However, measuring the results of energy *management* investments is different. Some projects may only require a measurement of a performance indicator. But the value of most projects is based on the concept of reducing consumption over time. This reduction is often called "savings."

Measurement and verification (M&V) is the process of determining the energy and value impacts of energy management activities. The process begins with a general assessment of the overall situation and the formulation of an M&V plan acceptable to all parties. It ends with agreement among all stakeholders on the reported results of the M&V activities. In M&V, the results of energy management projects are determined through calculations, using data gathered in the field. The calculations used to quantify results require models. Yes, there are measurements in M&V. And yes, verification plays a role. But perhaps the most important element in M&V is the model that represents the behavior of the existing system. In truth, a better name for M&V would be M, M, & V – measuring, modeling, and verification.

[1]https://www.unep.org/explore-topics/energy/why-does-energy-matter

Whether we call the results "savings," "impacts," or some other term, the outcomes of energy management are calculated. They cannot be directly measured. Hence, the exact method of calculation – and the data used in calculations – will impact the "results." In other words, there is no independent source of "truth" about savings. The only truth in M&V is the proper execution of an agreed-upon M&V plan. And without professional review, there is no guarantee that any particular M&V plan is sound. Some plans are constructed and executed better than others.

This book illuminates the role of the certified measurement and verification professional in imagining, developing, implementing, and reporting successful M&V projects.

While a basic understanding of energy management is a prerequisite for a successful M&V professional, it is equally critical that the M&V professional be aware of the context. Furthermore, experience has shown that the *primary role of the M&V professional* is to facilitate clear, terminologically consistent communication among all parties involved in the M&V process. This often requires the M&V professional to be aware of the broader financial, contractual, and regulatory context in addition to fundamental engineering and technical concepts. Hence, this book, while covering many technical topics, will focus on the broader issue of how an M&V professional can effectively apply fundamental M&V *concepts* in a wide range of energy management program *contexts*.

M&V professionals are called upon to make sound professional judgments. While M&V is grounded in the fundamental laws of physics, the process of applying M&V often requires many decisions to be taken on uncertain grounds. For instance, in addition to understanding the legal and regulatory situation, the M&V professional must often decide how and where to spend limited M&V resources. Choosing a less expensive flow meter may lead to different measurements of thermal energy (or more uncertainty). Selecting a smaller sample size will lead to greater uncertainty in the reported parameter. These are realities that must be addressed when applying M&V principles in any project. The M&V professional must be willing and able to make decisions that will ensure that the results reported from the M&V process are accepted by all.

This book is intended to provide the reader with a broad understanding of the role of the M&V professional in implementing best-in-class M&V. It will not delve too far into details about analysis techniques, metering tools, contract law, or rate design. Resources exist to address every situation, analytical and physical. But these have little value unless stakeholders can agree on how they will conduct M&V on their projects.

The range of skills required of the "complete" M&V professional is enormous. Most likely, no one person will be able to master them all. An M&V professional is someone who knows how to build a team to accomplish the *successful settlement* of a specific project in a particular context – an open-minded generalist who understands the core constraints and challenges.

Acknowledgments

I would like to express my gratitude to the many individuals who contributed to the creation of this book on Measurement and Verification. While it's impossible to name everyone individually, I want to extend a heartfelt thank you to some key contributors along the way.

My sincere appreciation goes to Eric Piller, whose editorial expertise and dedication played a crucial role in refining and enhancing the content of this manuscript. His insightful feedback and attention to detail have greatly contributed to the overall quality of the book.

Dale Sartor at LBL gave me my initial ticket to ride. Steve Schiller did most of the driving. Mark Ginsberg supported the early FEMP activities. Dr Jeff Haberl and the folks along the Brazos kept us honest. Greg Katz threw gas on the fire. Satish Kumar managed the ship once we left port.

Doug Dahle, Art Rosenfeld and Shirley Hansen were crucial supporters of M&V and legends in the industry.

Dozens of volunteers over two decades at the Efficiency Valuation Organization have kept M&V a "thing". And that's good for the planet.

The entire team at the Association of Energy Engineers and the dedicated group who support the CMVP are assuring M&V stays relevant as the energy transition progresses.

How lucky was I to have the team of Chris Balbach, Hassan Younes, Christo Van de Merwe, Antonio Miranda, Jon Feldman, Sandeep Dahiya, David Jump, Sam Day and Anna Kelly to work on the CMVP job task analysis and materials update? Very.

In the spirit of "If you want to learn something, teach it", I thank all the students who I've had the honor of learning from.

Finally, Carmen Best, thanks for enduring (and curtailing) my circumlocution. You all have much to thank her for!

To all those who contributed, directly or indirectly, your influence is imprinted on these pages. Thank you for being an integral part of this journey.

Best wishes,
Steve Kromer

List of Figures

List of Tables

List of Abbreviations

ACH	Air changes per hour
AEE	Association of Energy Engineers
AESP	Association of Energy Service Professionals
AFDD	Automated fault detection and diagnostics
AIA	American Institute of Architects
AIC	Akaike information criterion
AMI	Advanced metering infrastructure
AMR	Automated meter reading
ANOVA	Analysis of variance
ASHRAE	American Society of Heating, Refrigerating, and Air-Conditioning Engineers
ASME	American Society of Mechanical Engineers
ASPE	American Society of Plumbing Engineers
BEIS	Building energy information system
BEMCS	Building energy management and control system
BIC	Bayesian information criterion
BOMA	Building Owners and Managers Association
BREEAM	Building Research Establishment Environmental Assessment Method
BTU	British thermal units
CDD	Cooling degree days
CEM	Certified Energy Manager
CEN/CENELEC	European Committee for Standardization and the European Committee for Electrotechnical Standardization
cfm	Cubic feet per minute
CRI	Color rendering index
CSR	Corporate social responsibility
CT	Current transformer
CV	Coefficient of variation
CV(RMSE)	Coefficient of variation of RMSE
DER	Distributed energy resources

DMM	Digital multimeters
DOD	Department of Defense
DOE	Department of Energy
DR	Demand reduction
DRAS	Demand response automation server
DSM	Demand-side management
EAAS	Efficiency as a Service
ECAM	Energy Charting and Metrics
ECIP	Energy Conservation Investment Program
ECM	Energy conservation measure
EDM	Energy Data Management
EE	Energy efficiency
EEM	Energy efficiency measure
EEMS	Enterprise energy management system
EEP	Energy efficiency project
EIA	Energy Information Agency
EMV	Evaluation, measurement, and verification
EMIS	Energy management information system
EnBs	Energy baselines
EnPI	Energy performance indicator
EPA	Environmental Protection Agency
EPACT	Energy Policy Act
EPBD	Energy Performance of Buildings Directive
EPC	Energy performance contracting
ESAP	Energy Services Acquisition Program
ESCO	Energy services company
ESM	Energy savings measure
ESP	Energy service provider
ESPC	Energy savings performance contract
ESTCP	Environmental Security Technology Certification Program
EUI	Energy use index
EUL	Effective useful life
EV	Electric vehicle
EVO	Efficiency Valuation Organization
FEMP	Federal Emergency Management Program
FIFO	First in, first out
FSGIM	Facility smart grid information model
GBC	Green button connect

GBI	Green Building Initiative
GDP	Gross domestic product
GEN	Global ESCO Network
GHG	Greenhouse gas
GUM	Guide to Uncertainty in Measurement
HDD	Heating degree days
HVAC	Heating, ventilation, and air conditioning
ICC	International Code Council
IECC	International Energy Conservation Code
IEPEC	International Energy Program Evaluation Conference
IES	Illuminating Engineering Society
IIoT	Industrial Internet of Things
IMT	Inverse Modeling Toolkit
IoT	Internet of Things
IPMVP	International Performance Measurement and Verification Protocol
IR	Infrared
ISO	International Standards Organization
ISP	Industry standard practice
KPI	Key performance indicator
kWh	Kilowatt-hours
LBNL	Lawrence Berkeley National Laboratory
LEED	Leadership in Energy and Environmental Design
LINE	Linear regression interactive estimation
lpm	Liters per minute
M&T	Monitoring and Targeting
M&V	Measurement and Verification
MCMC	Markov chain Monte Carlo
MSE	Mean squared error
MUA	Measurement uncertainty analysis
NASEO	National Association of State Energy Officials
NBI	New Building Institute
NEB	Non-energy benefit
NEC	National Electrical Code
NEMA	National Electrical Manufacturers Association
NEMVP	North American Energy Measurement and Verification Protocol
NFPA	National Fire Protection Association

NIBS	National Institute of Building Sciences
NIST	National Institute of Standards and Technology
NMEC	Normalized metered energy consumption
NRE	Non-routine event
NREL	National Renewable Energy Laboratory
NTG	Net-to-gross
O&M	Operations and maintenance
OSHA	Occupational Safety and Health Administration
PV	Photovoltaic
RCT	Randomized control trial
RCx	Retrocommissioning
RMSE	Root mean squared error
RPS	Renewable portfolio standards
RTD	Resistance temperature detectors
RUL	Remaining useful life
SCADA	Supervisory control and data acquisition system
SEM	Strategic energy management
TC	Technical Committee
TMY	Typical meteorological year
TOU	Time-of-use
TOWT	Time-of-Week and Temperature
TRM	Technical resource manual
UMP	Uniform Methods Project
UNEP	United Nations Environment Programme
USGBC	U.S. Green Building Council
UT	Universal translator
VEAT	Virtual energy audit tool
VOI	Value of information
VPI	Value of perfect information

1

Introduction

M&V is the accounting system of energy management.

The aim of M&V is to produce credible estimates of the impacts of energy management projects, with the ultimate objective being the successful *settlement* of these results, agreed upon by all project stakeholders.

While the basic concept of improving performance is simple, energy projects' engineering, economics, and legal issues can often be very complex. Efforts to develop a simple set of solutions to the problem of quantifying the impacts of energy management activities have met with serious challenges due to the breadth of applications and the scale of the industry. Accordingly, the guidelines, methods, and even terminology of M&V can vary widely from one context to another. This book provides an overview of the range of applications that an M&V professional may encounter and the fundamental principles underlying all M&V activities.

The fundamental concepts of M&V are elementary. They are not unique to this profession, but rather derived from basic disciplines, such as science and econometrics. Added to this foundation are general concepts like lifecycle costing and utility tariff/rate analysis. But while the ideas are simple, their implementation can often be extremely complex. As an M&V professional, you are not expected to be an expert in all the details, from engineering to statistical analysis to finances. Instead, you will most likely lead a team in which individual members contribute specific knowledge that collectively informs the M&V plan and process. Your primary goal is to ensure that the M&V activities undertaken generate credible estimates of impacts and ensure that all stakeholders understand and agree on how the M&V process will unfold. Yes, M&V involves plenty of technical tasks, such as baselining, metering, and analysis – but behind this veneer of objectivity, the truth is that you must exercise professional judgment, make tradeoffs, and document your decisions in a manner acceptable to all stakeholders. In fact, the most important tool in achieving your goal is to establish and maintain *clear and consistent communication* among all relevant stakeholders throughout the M&V process.

And then there is the ever-present reality of uncertainty.

In M&V, uncertainty refers to our degree of confidence in the accuracy and reliability of the data collected and analyzed. Uncertainty can arise due to factors within and outside the control of the M&V professional. Identifying sources of uncertainty, quantifying them, and managing their influence are a critical component of the work of M&V.

Where does an M&V professional begin? The first step is to understand the energy management activity's nature and context. We will frequently use the terms *concepts* and *contexts* to emphasize that M&V activities apply to a plethora of energy management projects and programs.

1.1 Who This Book Is For

"*M&V is too important to be left to the engineers.*" – Steve Kromer, P.E.

I wrote this book with three specific audiences in mind:

- It will be of interest to *those involved in decision-making about the application of M&V*. If you are involved in strategizing or implementing M&V, you will find this book a helpful companion.
- The book is also for *those hoping for an introduction* to M&V. Whether you are a student, a new practitioner, or simply curious about the profession, this book provides a clear, comprehensive view of what M&V is – and what it is not.
- Finally, the book caters to *experienced practitioners who would like to learn more* about the core theories of M&V and the tradeoffs that can arise when using M&V. This book will give you a better grasp of the complexities and considerations involved in these procedures.

1.2 The Structure of This Book

This book consists of three major sections: *Concepts and Contexts*, *Standards of Practice*, and *Practical Considerations*. The first of these sections, *Concepts and Contexts*, will begin by familiarizing you – or, in some cases, refamiliarizing you – with the fundamental concepts of M&V, from its place within energy management to the relationships between cost, risk, and benefit. The section will also help you more fully contextualize the work of

M&V, which not only involves engineering but also extends into contractual, regulatory, and financial domains. Finally, *Concepts and Contexts* will discuss the characteristic activities of the M&V professional – while also defining what exactly it means to be an M&V *professional.*

The next section of this book, *Standards of Practice*, considers the role of standards, protocols, and other guiding documents in M&V. Through this lens, it briefly recounts the history of M&V and offers an insider's perspective on some of the decisions that ultimately shaped the profession.

The third major section, *Practical Considerations*, takes a deeper dive into many of the concepts introduced earlier in the book, such as modeling, metering, and valuation. It provides a wealth of practical guidance for the M&V professional, gleaned from my 30 years of experience in the profession. Each of the subchapters in this section could fill another book. For now, I hope to create awareness of the issues you will face and resources for further investigation.

2

Concepts and Contexts – The Activities of the M&V Professional

In your duties as an M&V professional, you will be frequently tasked to explain the art and science of M&V to a wide range of participants. Experience has shown that you will find that many people have no idea how M&V "works." Some may think it is a set of exact techniques that can generate precise answers. Others may see it as smoke and mirrors – much ado about nothing. And in some ways, both are right. The tools and methods can generate impressive-looking data that represents the "absence" of energy. But M&V is really a set of concepts that allow us to make reliable assessments of impacts of energy management – if everyone is familiar with the process. In Chapter 3, we will look at some of the standard documents that have come to define M&V for many people. But first, we will explain how a very simple concept is the foundation of all the complexities addressed in those documents. We will also see how it is critical for the M&V professional to appreciate that the same simple concepts can be applied in different ways depending on the context. We intend this to be a firm foundation for the details that follow in Chapter 4.

2.1 Concepts

2.1.1 What is measurement and verification?

The first challenge facing the M&V professional is to understand the wide range of definitions applied to the term "measurement and verification" and the shorter "M&V." It is common to hear "M&V" used as shorthand for "monitoring and verification" or a range of other phrases such as "evaluation, measurement, and verification" (EM&V). You will hear EM&V when dealing with the evaluation of utility energy management programs.

Today, in practice, M&V is a set of concepts meant to address the needs of the energy management industry for the quantification of the impact of energy management projects. For the purposes of this book, we will consider it to be a *process* through which project or program stakeholders envision, develop, implement, and report (i.e., settle) an M&V plan or procedure, whether it be site-specific or applied to a portfolio of projects. This is a broad definition, reflecting common usage, and encompasses both long-term "savings" and performance verification.

In an exercise on LinkedIn in 2021, I offered a bounty to anyone who could definitively identify the first use of "measurement and verification" in its current sense. A winner has not yet been declared, but the best information provided so far indicates that the term "M&V" originated in the early 1990s. The following is a rough timeline of the development of guiding documents.

***

- 1970s: Case-by-case measurements
- 1983: International Energy Agency's "Guiding Principles for Measurement"
- 1985: Oak Ridge National Laboratory's "Field Data Acquisition for Building and Equipment Energy Use Monitoring"
- 1988: New Jersey Utilities M&V Guidelines
- 1988: First NAESCO M&V Guidelines
- 1989: Texas LoanSTAR Program M&V Guidelines
- 1991: ASHRAE Handbooks Chapter 37 "Building Energy Monitoring"
- 1992: California CPUC CADMAC M&V Protocol
- 1993: New England AEE M&V Protocol
- 1993: NAESCO M&V Guideline ver. 1.3
- 1994: PG&E PowerSaving Partners "Blue Book"
- 1995: EPA Conservation Verification Protocols
- 1996: NEMVP
- 1996: IPMVP
- 1996: FEMP M&V Guideline ver. 1 (currently ver. 4, 2015)
- 2002: ASHRAE Guideline 14 ver. 1 (currently ver. 3, 2022)
- 2016: IPMVP Core Concepts (currently EVO 10000 – 1:2022)

***

Note that until the early 1990s, the energy management industry was still defining "M&V" as field monitoring and measurement. Throughout this book, we will attempt to distinguish the concepts of field measurement, measurement of performance, and calculations (determination) of accrued impacts. And in Chapter 3 we will discuss the evolution of M&V from it's beginnings in field measurement to savings determination.

The first national guidance document for M&V was the 1996 *M&V Guidelines*, issued by the U.S. Department of Energy's (DOE) Federal Emergency Management Program (FEMP). Soon after, the DOE released the predecessor of the International Performance Measurement and Verification Protocol (IPMVP): the North American Energy Measurement and Verification Protocol (NEMVP).

According to FEMP's guidance, a site-specific M&V plan was required for every project. IPMVP also defines M&V in terms of creating a site-specific M&V plan for each project. However, in the intervening decades, the general term "M&V" has been applied to projects and programs where there is no site-specific plan. For example, FEMP/IPMVP "Option C" became synonymous with the "utility bill" or the "whole facility" approach to M&V. This approach has become common in "pay for performance" programs where the participating projects do not receive an M&V plan.

This is one of the first and most important lessons for the M&V professional – to develop among the stakeholders a common understanding of the definition of "M&V."

Throughout this book, we will emphasize the need for an M&V professional to be alert to the fact that there is a wide range and variety of terms used in the practice of M&V. We will attempt to distinguish general terminology from specific definitions.

Some examples of terminology used in utility program evaluation, measurement, and verification (EM&V), but not typically seen in energy savings performance contracts (ESPC), are words like "ex-ante," "ex-post," and "realization rate." These concepts are defined in standard guidance documents that serve a specific context, such as the Californian Evaluation Framework and the Strategic Energy Management M&V Protocols. An M&V professional must understand the core concepts that underlie all these documents but be prepared to adapt to the methods and definitions used in each context.

One thing is for certain: practicing M&V always includes some form of data collection. It also involves:

- determining what measurements will be made, where, for how long, and at what cost;
- how the data will be analyzed and adjusted;
- how the results of the study will be reported.

The goal of these activities is to reach a settlement – the execution of payments that account for the impacts or "results" of the energy management project. I prefer "impact" over the more common "savings," because "savings" is meaningless outside of a specific context and without additional information about assumptions, as we will discuss later in this chapter.

2.1.2 Energy management methods and types

Energy management can be broadly classified into four principal activities: equipment upgrades, behavior/set points, demand management/flexibility, and fuel switching.

- **Equipment upgrades:** This activity entails replacing or upgrading equipment to increase efficiency, which can reduce energy consumption and associated costs. Examples include replacing old lighting fixtures with energy-efficient LED bulbs and upgrading HVAC systems.
- **Behavior/set points/controls:** This activity involves changing human behavior or adjusting set points to reduce energy consumption. Examples include turning off lights or equipment when not in use and adjusting thermostat settings to reduce heating or cooling when a building is unoccupied.
- **Demand management/flexibility:** The crux of this activity is to reduce energy consumption during peak demand periods, often through demand-response programs. For example, a facility might reduce energy consumption during periods of high electricity demand by adjusting equipment schedules or temporarily shutting down non-essential equipment.
- **Fuel switching:** This is simply to switch to a more efficient and/or cost-effective fuel source. Examples include switching from oil to natural gas for heating and using renewable energy sources like solar or wind power rather than non-renewable fuels.

Each of these activities can be part of an energy management program. The best approach – or a combination of approaches – will depend on the facility's energy usage patterns, the project or program's goals, and available resources. An energy audit can identify opportunities for improvement in each of these areas and help guide development of an energy management plan.

Whatever the type of intervention, stakeholders will universally desire to determine if the project is successful, and success is typically quantified in terms of a reduction in energy consumed. But even that simple definition does not apply to some current programs. Sometimes the goal of energy management is to shift loads rather than reduce them. We will attempt to give the M&V professional the widest view of current and evolving concepts in M&V.

As an M&V professional, you need to consider which types of energy management activities have been implemented and draft the appropriate M&V plan. Energy management activities such as behavior/set points and demand management/flexibility often require continuous metering. This is because the impact of these activities is based on relatively inconsistent factors: are users of the facility consistently turning off lights and shutting down non-essential equipment during peak demand hours? If you measure a facility's energy use only periodically, you might find yourself unable to answer these questions with a sufficient degree of confidence. In contrast, equipment upgrades and fuel-switching activities lend themselves better to periodic metering, as the performance improvements resulting from these activities are typically more "fixed." That said, even for equipment upgrades and fuel switching, it may still be worthwhile to have some continuous metering to ensure that the equipment is operating as intended and to identify any performance issues that may arise over time.

For programs that promote flexibility of loads, M&V will need to be customized to accurately measure the impact of load shifting or demand response activities on energy consumption. In such programs, customers may be incentivized to reduce or shift their energy usage during peak demand periods or in response to grid signals.

M&V for these programs may involve continuous monitoring of energy consumption before, during, and after load-shifting events to accurately determine the impact on energy usage. This may require the use of advanced metering systems that can provide real-time energy consumption data.

2.1.3 The common denominator problem

Every energy-consuming system exists for some reason other than just to consume energy. Every facility was designed to meet the needs of its users, investors, or other stakeholders – to provide light, create a healthy and comfortable environment, facilitate communications, manufacture widgets, etc.

In addition, every energy-consuming system operates within a matrix of physical, financial, and regulatory requirements and constraints. Some of these constraints are rooted in the purpose of the facility – hospitals, for example, must achieve a minimum amount of air changes per hour (ACH) for certain rooms, which might otherwise pose a danger to patients with heart and lung diseases or respiratory issues, and facilities where painting occurs must comply with OSHA requirements regarding ventilation. Other constraints arise from broader government requirements, such as energy codes.

Some of the requirements for and constraints on residential and commercial facilities relate to indoor air quality, lighting levels, comfort, appliance standards, and the suitability of the facility for its intended purpose. For industrial facilities, they often have to do with production levels, safety requirements (ventilation and adequate lighting), availability of fuels (coal and gas), and so on.

The challenge is to understand the needs and preferences of the energy system of interest. In some cases, an absolute reduction in energy use is desired and justified. In other cases, if energy productivity can be increased, facility owners may wish to use more energy – with the benefit of enhanced profitability in mind. In some cases, regulations will require the facility to meet a target. In others, the most important metric may be sales volume per square foot, requiring high intensity lighting. The bottom line (the denominator) is that you should consider service levels and productivity, not just raw energy inputs.

Note:

"Productivity" can be difficult to measure, especially for office buildings, etc. – adds complexity to common denominator problem.

When an energy management project is implemented, it is expected that the facility's service level and its compliance with relevant codes and standards will be achieved once the project has concluded. But what if the facility's baseline conditions were not meeting code in the first place? What if the result of the energy management project is *increased* consumption to address insufficient service levels – whether they be environmental conditions in a residential or commercial building or some productivity metric in a manufacturing facility? This calls into question what constitutes a fair and relevant baseline.

Here, as an M&V professional, your job becomes more complicated. You must consider how to create an M&V plan that all stakeholders can accept, when it is likely that the result of the project is increased energy consumption. You must communicate the concept that the only fair and relevant baseline is one where the facility meets all codes, regulations, and service requirements.

This sounds like an extreme case, though in practice it is common in some utility programs such as commissioning. A facility's existing equipment may be near the end of its useful life, or the facility and/or its equipment may not meet current energy codes. Still, even when the facility *is* operating to code, you need to understand the entire context of the existing facility, including the current stage of the facility's lifecycle and the relevant codes, standards, and market conditions within which the energy management activities will occur. As an M&V professional, you must work with stakeholders to ensure a shared understanding of:

- the status of the existing facility, not only in terms of energy efficiency but also usefulness, productivity, and any other relevant aspects;
- the nature of the intervention and expected energy consumption and service levels;
- the equipment to be replaced, remaining useful life, and ongoing maintenance costs;
- the tool (models and calculations) that will be used to adjust for future conditions.

All of this requires some degree of judgment.

Finally, our recent global experience with the COVID-19 virus led to changes in outdoor air requirements for many buildings. This new service level requirement requires new accounting for the baseline energy use.

* * *

Replacing underperforming equipment may be warranted, but the attribution of credit for the "savings" can be a complicated matter. In Chapter 4, we will discuss dual baselines and other relevant concepts. For now, consider these examples:

- A 35-year-old boiler is replaced by a new boiler. Market transformation has occurred, and inefficient boilers like the one previously in the facility are no longer available. What is the proper baseline?
- A gas furnace is replaced by an electric heat pump. Gas appliances are outlawed in the facility's jurisdiction. What is the proper baseline?

* * *

2.1.4 Lifecycle thinking

Every energy-consuming facility exists in a particular stage of its lifecycle. This lifecycle spans various phases:

- **Design:** At this stage, architects and engineers develop plans for the facility's structure and systems. This is when key decisions are made that will impact the building's energy efficiency, including choices about building materials, orientation, insulation, HVAC systems, and more.
- **Construction:** Here, contractors and other professionals turn the design plans into reality, constructing the facility and installing its systems.
- **Commissioning tests:** At this stage, the newly installed systems are tested to ensure that they are functioning as intended. Adjustments may also be made as needed to optimize performance.
- **Operation:** During this phase, the facility is in use and its systems are functioning on a regular basis. Ongoing energy management activities, such as monitoring energy usage and conducting regular maintenance, are key to sustaining the building's energy efficiency over time.
- **Retrofitting/renovation:** Energy retrofits may be considered when the facility's energy performance begins to decline. Or, if new, more efficient technologies become available, it may be beneficial to retrofit or renovate the facility's systems by updating or replacing its equipment. Renovations and retrofits could also happen due to changes of space type or many other reasons that might lead to an increase or decrease of energy consumption and/or demand.
- **Teardown:** This is the final stage of a facility's lifecycle, when it is decommissioned, and its components are salvaged for reuse or recycling. Even at this stage, considerations related to energy and sustainability can play a role – for example, by choosing deconstruction methods that allow for maximum reuse of materials or by recycling materials in an energy-efficient way.

As an M&V professional working on a particular project, you should seek to understand both the original design intent of the facility and what has occurred over its life such that there are now opportunities for energy management. Have operating requirements changed? Is new technology available to replace existing less efficient equipment? Understanding the underlying changes will help you craft the M&V plan.

For example, there are still many buildings using lighting systems designed decades ago. Changes in lighting equipment, without changing lighting levels or operating hours, can often be verified with simple methods. Conversely, replacing a constant speed motor with a variable speed drive would require more data on operating characteristics.

2.1.5 Jevon's paradox

An additional issue to consider is the rebound effect, also known as Jevon's paradox. This principle states that as technological improvements increase the efficiency of resource use, the rate of consumption of that resource may increase as well, contrary to expectations. For example, if a factory upgrades to more energy-efficient machinery, the facility may ultimately consume more energy overall if the cost savings allow it to ramp up production. Jevon's paradox complicates the work of the M&V professional by challenging the conventional understanding of energy conservation and efficiency measures. It is a caveat that must always be kept in mind when projecting the impacts of energy efficiency measures, and it points to the importance of establishing a clear definition of "success."

Example: In a lighting energy retrofit project, the occupants kept the lights running 24/7 because they thought the new lights had "minimal" electricity consumption – as the ESCO discovered, when re-lamping was required much earlier than anticipated.

2.1.6 Energy management – and the need for accountability

Energy management comes in different forms. M&V supports energy management by providing a logical framework within which all parties can define success (savings). Table 2.1 lists some of the common terms used for energy management projects.

Energy Management Terminology

Energy management includes several concepts and activities: energy conservation, efficiency, productivity, flexibility, optimization, and management.

- Both *conservation* and *efficiency* aim to reduce energy consumption, but they differ in their approaches. Conservation focuses on reducing overall energy use, while efficiency focuses on maximizing the output per unit of energy input. For example, turning off lights when not in use is a conservation measure, while replacing incandescent light bulbs with LED bulbs is an efficiency measure.
- *Productivity* focuses on maximizing output per unit of energy input. This means that productivity measures may not necessarily reduce energy consumption but can still lead to increased efficiency. For example, optimizing a manufacturing process to reduce the amount of energy required per unit of production can increase productivity.

Table 2.1 Terms used for energy management interventions.

Energy efficiency	EEM (energy efficiency measure)
Energy conservation	ECM (energy conservation measure) ESM (energy savings measure)
Energy performance/productivity	EnPI (energy performance indicator)[1]
Flexibility	DERs (distributed energy resources)[2]
Energy optimization	EE (energy efficiency), DER, fuel switch[3]

- *Energy flexibility* is a relatively new term. "Flex markets" employ a range of technologies that allow the demand side to adapt to the changing energy technologies (also known as DERs). In other words, flex markets promote the use of energy when it is most readily available and cost-effective and, in this way, reduce strain on the energy grid during peak demand periods.
- *Optimization*, finally, refers to how these various activities work together to improve overall energy performance.

What is the Real "Product" of Energy Management?

Hint: It is not the equipment that is installed.

The collective actions that make up energy management – changing out equipment, improving operations, etc. – can all be taken outside of the constructs of energy savings performance contracts or energy efficiency programs. In many situations, a simple estimate of "return on investment" is sufficient. The cost is known. The value is estimated.

In these cases, the evidence that the new equipment is installed and working properly is enough to justify paying the invoice. And in these cases, M&V is rare.

As we will see in Chapter 3, M&V blossomed as a professional practice when the "product" of energy management projects changed from the equipment installed to the quantification of the reduction in payments necessary to operate a facility. Traditionally, the word used to describe the product or outcome of energy management activities is "savings."

This is unfortunate. "Savings" is insufficient as a concept. It is overly simplistic and easily misunderstood. The main problem with the use of the word "savings" is that it suggests that a reduction in energy consumption

[1]https://www1.eere.energy.gov/manufacturing/eguide/iso_step_2_7.html

and cost can be unequivocally assessed. But without an agreed method for quantification, it is subject to interpretation and hence disagreements.

But, some will say, the basic process of M&V is to collect data on energy consumption before and after the implementation of one or more energy management measures and compare the data to determine the actual energy savings achieved.

Simple, right?

Simple in theory, yes. But the reality is that facilities do not remain stable over time in a way that would allow for a simple pre/post comparison. Future conditions will differ from the baseline conditions. Changes are made to the facility. M&V is the accounting system that tracks these types of changes to determine estimated "impacts."

Certain characteristics of energy performance can be measured and verified, but "savings" – effectively the absence of energy – must be derived through calculations.

Defining what calculations will be used – before the project is installed – helps ensure that all stakeholders have a clear understanding of what is being measured and how the impacts will be calculated. The goal is to prevent misunderstandings, maintain transparency, and create credibility in the results.

Let us all say it together, loudly: "Savings are defined, not measured." Wherever possible throughout this book, we will use the word "impacts" to maintain this point.

Agreed-upon impacts are the goal of energy management. Their quantification is the goal of M&V.

To reiterate, the energy management concept might be to install a new chiller. That new chiller may be installed and running. It may look great. But that is not sufficient to prove that it is "working." The only way to prove that it is delivering the promised impacts is to have baseline data on the old system that is adjusted to the reporting period conditions. The product of energy management in this case is not the equipment – it is the stream of reduced costs resulting from that equipment.

What is M&V? It Depends

The main protocols and guidance documents provide for a wide range of activities that can be called M&V. One extreme might be a detailed scientific study to evaluate a utility program. The other extreme might be a simple performance check to verify that a piece of equipment is working properly.

This range of M&V approaches can be generalized into two main categories: performance verification and impact ("savings").

Performance and impact are related but distinct concepts. Briefly, performance is a measure of how well a system or process is functioning, while impact is a measure of the operational benefits that result (and accrue) over time from improved performance. To elaborate:

- Performance refers to how well a system or process is functioning, or how well it is achieving its intended goals. It can be measured in a variety of ways, depending on the specific system or process. It is always stated as a characteristic of a system.
- Impact, on the other hand, refers to the reduction in costs or resource consumption over time that results from an improvement in performance. It can be quantified in terms of financial impact, such as lower energy costs, or in terms of resource impact, such as reductions in CO_2 emissions. Impact – or "savings" – is stated as an amount.

How do these concepts translate into the work of an M&V professional? Performance can be measured. Impact requires a model, informed by measurements. It focuses on establishing a means of comparing the actual energy use to an agreed-upon amount of energy that "would have been" consumed.

Each approach has its own benefits and limitations. The choice of which approach to take will depend on the specific project goals and circumstances. Clearly explaining the differences between these approaches – and using precise and clear language *throughout* the M&V process – can help ensure that all stakeholders have a sufficient understanding of the results.

2.1.7 The theoretical basis of M&V

> "Mr. Kromer is here to discuss M&V, which is just the scientific method using different words." – Indian Energy Minister's introduction to the first IPMVP Seminar in Delhi, October 2005

In Chapter 3, we will discuss the evolution of M&V protocols – specifically, how organizations came together to establish an industry consensus on standard methods for quantifying the impacts of energy management. As with any work that involves committees, the results – the printed documents – represent a balance of the opinions and perspectives of the authors. But they are all grounded in fundamental scientific and engineering principles, while also recognizing that any resulting methods must be realistic to apply in real-life business settings.

M&V as a Scientific Pursuit

M&V's credibility is rooted in its application of the scientific method. By taking a scientific approach, you can establish clear and transparent methodologies for quantifying the impact of energy management activities, which helps ensure that the results are trustworthy and can be replicated by others. The use of the scientific method also minimizes the effect of bias or other factors that could potentially compromise the accuracy and validity of the results.

Much scientific research is based on hypothesis testing, a statistical method used to test a hypothesis or claim about a population based on sample data. In hypothesis testing, a threshold of evidence is established, typically a p-value, which represents the probability of observing the sample results or more extreme results if the null hypothesis (the hypothesis that there is no effect or difference) is true. If the p-value is less than a predetermined level of significance, typically 0.05, then the null hypothesis is rejected, and the alternative hypothesis is accepted. In other words, a hypothesis test can determine a yes or no answer to whether there is evidence of an effect or difference.

But research methods can also seek to determine the quantitative impacts of interventions: *impact analysis*. Rather than applying the null hypothesis, impact analysis is used in M&V to quantify the expected impact of an energy management project with an associated confidence interval.

This type of analysis can be done using a variety of methods, such as simulation models, statistical models, and data analysis.

In this sense, M&V can be thought of as simply *a set of specific terms used to adapt fundamental scientific principles for use in the context of energy management*. The general concepts of M&V are like those of any other scientific pursuit. The primary difference is this: applying the scientific method requires that all variables – independent variables and "static factors" – affecting the outcome of the experiment must be identified and accounted for. This is often impossible in energy management projects; so M&V must also allow for cost-effective, "good enough" answers.

While scientific methods are core to M&V, there are also non-scientific elements at play. For example, there may be legal or regulatory requirements that must be followed, such as complying with building codes or appliance standards. Specifics about the actual operation of a facility – the answers to questions like "who turned out the lights?" – can matter a great deal as well. There can be significant financial risks involved, such as fluctuations in energy prices or unexpected project costs. And, of course, M&V must be performed within a project's financial constraints: a more expensive, more elaborate, and ultimately more precise approach may be unwarranted if the higher level of certainty that it brings holds less value than the added cost. Thus, while a scientific approach is key to the work of M&V, it must be considered in tandem with effective risk management strategies, compliance with legal (e.g., contractual) and regulatory requirements, and other factors necessary to a project's overall success.

To put it differently: is M&V energy engineering, econometrics, or a scientific pursuit? The simple answer is "all of the above." As a scientific pursuit, the work of M&V requires an understanding of the physical principles that govern energy systems. As energy engineering, it requires an understanding of the technical details of energy systems, including equipment specifications, operating conditions, and maintenance practices. And as econometrics, it requires an understanding of the financial implications and constraints of energy management projects. As an M&V professional, you are responsible for integrating the scientific, engineering, *and* financial aspects of M&V, while also accounting for the regulatory environment and relevant contractual or programmatic stipulations.

Scientific Terminology: Quasi-experimental vs. Experimental

In one, formal definition of an experiment, the researcher manipulates an independent variable to observe its effects on a dependent variable.[2] They then compare the results to a control group, which has not been exposed to the independent variable. Throughout this process, the researcher tries to eliminate or control all variables other than the independent variable, so that any observed effects are due to the manipulation of the independent variable.

A quasi-experimental approach is similar to an experiment, but it does not include a control group. Instead, the researcher compares the impact on the experimental group to a comparison group that is similar in all other respects except for the independent variable. This type of design is often used in situations where it is not feasible or ethical to have a control group, such as when studying the effects of a policy change. This method is often applied in utility programs.

M&V can be considered quasi-experimental because it (typically) involves a comparison of the performance of an energy system or building before and after a change is made, without a control group. Here,

[2]In the case of energy management projects, the independent variable is the project being implemented.

the "change" (or intervention) could be the installation of new equipment, changes in operating procedures, or other energy-saving measures. However, any M&V design must take into consideration the need to discriminate between the impact of the energy management project and other co-occurring events. As we will see, the M&V community has adopted specific terminology for these events (routine and non-routine).

In many situations, monitoring every potential factor is cost prohibitive. To reduce the cost of M&V to an acceptable level, M&V professionals often take pragmatic steps, such as sampling a subset of the energy system or facility rather than measuring everything. This allows for reasonable estimates of the energy impacts while minimizing the time and cost required for data collection and analysis. Additionally, statistical techniques can be used to estimate the impact of the energy management project with a quantifiable degree of confidence, even given a smaller sample size. However, the results of a quasi-experimental approach are not necessarily as strong as those of an experiment with a control group, and interpretation of the results must consider the limitations of the design.

Threats to Validity

"Threats to validity" are factors that can affect the accuracy and reliability of data and, by extension, the conclusions derived from a study. This concept is fundamental in research methodology, and it is particularly relevant to the work of M&V professionals. In the context of M&V, threats to validity can arise from a range of sources, such as faulty measurement tools, biased sampling methods, uncontrolled extraneous variables, and data errors or omissions. You must be aware of these threats and take steps to minimize their effect on your results.

For example, as an M&V professional, you will often apply statistical techniques (regression analysis) to account for variables that influence energy use. You can also implement quality control procedures so that data is collected and analyzed consistently across the lifespan of a project. Most importantly, you should communicate effectively with stakeholders to ensure that they have a clear understanding of the goals and methods of the M&V process, and you should work with them to identify and address any potential sources of bias or error.

Science and Engineering

Throughout this book, we make the point that M&V is a judgment-based application of scientific principles applied to engineering projects dealing with energy management. We compare the goals of M&V in different contexts – where stakeholders have different needs, perspectives, and budgets. We emphasize that the specific skills an M&V professional must possess focus first on the creative application of M&V concepts to each project. But perhaps more importantly, the M&V professional must communicate a range of concepts to a wide audience – many of whom may not possess a detailed understanding of many of the topics involved. This is particularly true when it comes to the contractual and regulatory constraints applied to engineering details.

2.1.8 Counterfactual methods – modeling vs. comparison

> *Two roads diverged in a yellow wood...* – Robert Frost, "The Road Not Taken"

The two main methods that comprise M&V are performance measurement and impact ("savings") analysis. While performance verification is intended to confirm that equipment is operating as designed, impact analysis calculates the impact of an energy management over time and is an implementation of the counterfactual method.

The counterfactual method requires comparing an outcome that did occur (actual energy use) against a hypothetical outcome that did not occur (what the facility *would have consumed* absent the project).[3] Counterfactuals come in two types, sometimes called *cross-sectional* and *longitudinal* (Figure 2.1).

Cross-sectional methods apply the counterfactual method by comparing a specific project to a comparison group (or control group).

Longitudinal methods compare an energy-using system to itself over time. This requires a model of the system (baseline) to provide a reference case against which the actual use can be compared.

The longitudinal form of the counterfactual method (modeling) is the core of most M&V protocols, though either method may be used, independently or together, especially in the context of utility programs.[4]

To perform a counterfactual analysis using a baseline model, the details of the baseline (including the baseline period) are specified along with the period over which the impact analysis will be conducted – the reporting period.

---

[3]https://www.youtube.com/watch?v=GTgZfCltMm8 "CausalImpact 1.2.1, Brodersen et al., Annals of Applied Statistics (2015). http://google.github.io/CausalImpact/"

[4]Population NMEC – https://www.cpuc.ca.gov/-/media/cpuc-website/files/legacyfiles/n/6442463694-nmec-rulebook2-0.pdf

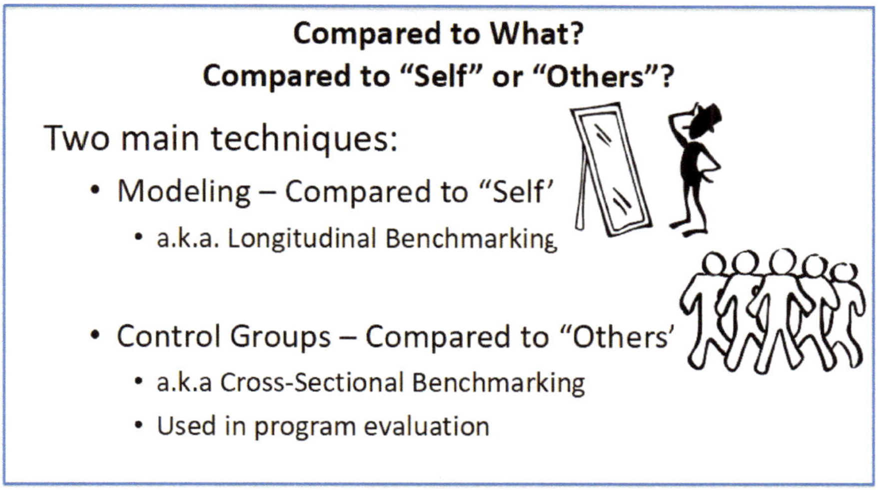

Figure 2.1 Two types of counterfactual methods.[5]

An effective baseline model accounts for the factors known to influence energy use or considered likely to influence energy use if they change during the reporting period (requiring adjustments) (Figure 2.2). This is sometimes called the "What **WOULD** Have Happened" meter.

A variant of the situation above is a counterfactual that is generated based on energy codes or standards. This counterfactual is sometimes called the "What **SHOULD** have happened" meter (Figure 2.3).

Later in this book, we will learn about specific types of judgment necessary when developing and adjusting baseline models. For now, let it suffice to say that the results of a counterfactual analysis are only as reliable as the baseline model, and the accuracy of the results will depend on the quality and completeness of the data and assumptions used to build the model.

Given that any specific baseline model will be the result of compromises and judgments (M&V is cost-constrained), the resulting target quantity – impact, or "savings" – will vary according to the details of the M&V project. Impacts are not directly measured; rather, they are always a function of the method applied. There is no such thing as an "accurate" impact calculation until the M&V approach is selected and the plan is defined.

[5]The terms cross-sectional and longitudinal benchmarking are from the Energy Information Handbook published by LBNL/DOE and available at: http://eis.lbl.gov

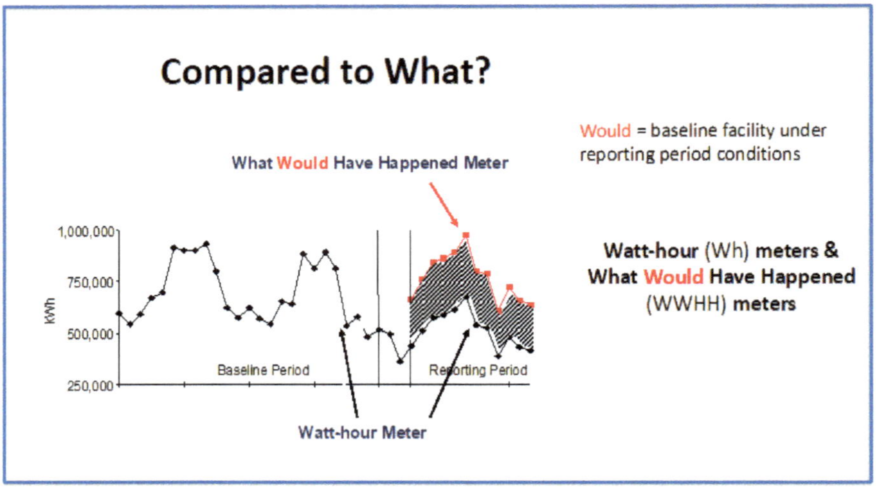

Figure 2.2 Counterfactual based on adjusted baseline.

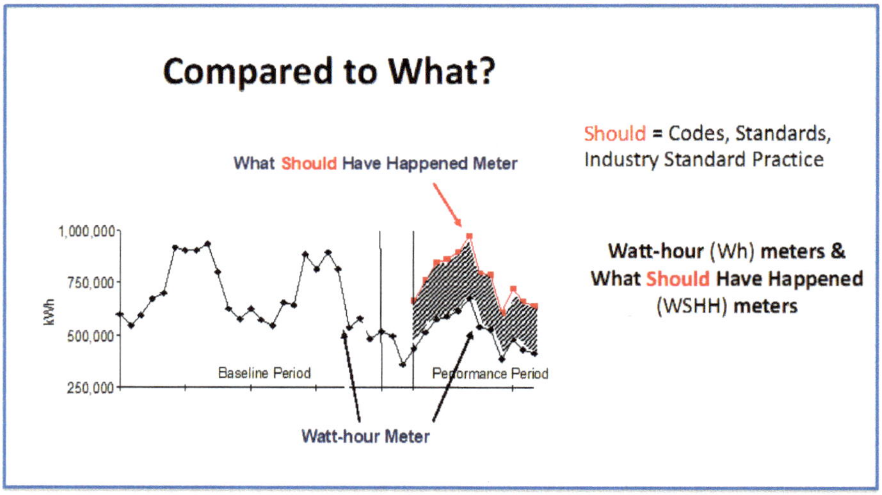

Figure 2.3 Counterfactual based on codes, standards, or industry standard practice.

One implication of this – and of the counterfactual method in general – is that there is a potential for gaming, i.e., manipulating a system or process to achieve a desired outcome. Because the counterfactual method utilizes a hypothetical (baseline) model, we cannot definitively say what would have occurred in the absence of a project. And whether intentionally or by accident, an M&V plan that incorrectly models the baseline may improperly increase (or decrease) the appearance of "savings" (impact).

Gaming the results in M&V can erode trust among the parties involved and undermine the credibility of the entire process, leading to a lack of confidence in the reported impacts and potential that parties resort to litigation. If one party believes that the other party is manipulating the results of M&V, it can lead to disputes, legal action, and damage to the reputation of the profession.

Over the years, there have been such situations. I coined the phrase "First Law of ESCO Dynamics" to playfully point out that, apart from ethical considerations, it is always in the interest of the party performing the energy efficiency project to inflate the baseline. To some extent, this can mean that the "savings" from the project are a negotiated quantity. While it is a core premise of the proper conduct of M&V that all parties should adopt the guiding principles, there is still a need for all parties to understand and agree on the implications of a specific M&V approach and plan.

Trust is essential when it comes to implementing energy management projects successfully and achieving the desired outcomes. It can be built through transparency, accountability, and the use of independent third-party M&V professionals, who can provide unbiased and objective verification of the project outcomes, reducing the potential for gaming and increasing trust among all parties.

* * *

Time present and time past
Are both perhaps present in time future,
And time future contained in time past.
If all time is eternally present
All time is unredeemable.
What might have been is an abstraction
Remaining a perpetual possibility
Only in a world of speculation.
What might have been and what has been
Point to one end, which is always present.
Footfalls echo in the memory
Down the passage which we did not take...

– T. S. Eliot, "Four Quartets"

2.1.9 Performance verification[6]

While the counterfactual method is the key to defining accumulated impacts over time (aka "savings"), performance verification includes a broad array of concepts and policies. Performance can imply anything from very short-term measurements of a lighting system to normalized annual consumption of a building (aka benchmarking).

Some typical M&V questions include: Is the system performing better than before the project was implemented? What does "performing better" mean (see inset)? How can one measure that? And how can one extrapolate from performance to long term impact?

Here again, language can be tricky. When dealing with energy consuming systems and energy management, we need to be able to distinguish between the inherent characteristics of the energy system/facility and the operational characteristics.

It is tempting to adopt a simple set of definitions for performance and operation. For example, a building disconnected from power will consist of a set of assets that have a potential to *perform*. Once the power is turned on, the building will be *operated* to perform desired functions.

But depending on the system of interest, performance and operation might be clearly distinguished or highly interdependent. For example, a lighting system performs when turned on – the operational consideration is the hours of use.

A more complex example is a chiller plant, the various controls of which influence both the performance and operation of the system.

And in the case of a building "performance" standard, the performance metric is typically the annual energy use of a facility. The goal is to achieve a target designated by a key performance indicator (KPI).[7]

As will be discussed in Chapter 3, many M&V documents provide for at least one method that does not involve a full, counterfactual baseline model and instead focuses largely on performance. In 1996, FEMP released the first edition of its *M&V Guidelines*, using an "Options" terminology that has since become standard. FEMP defined "Option A" as taking short-term performance measurements to assess the impact of an energy efficiency measure. The instantaneous or short-term performance can then be combined with (typically multiplied by) operational factors to estimate "savings."

---

[6]ASHRAE has initiated work on Buildiung Performance https://www.ashrae.org/technical-resources/standards-and-guidelines/titles-purposes-and-scopes

[7]https://www.energy.gov/eere/buildings/commercial-building-performance-monitoring-and-evaluation

Table 2.2 Performance and operational metrics for common systems.

Performance metric	**Operational metric**
Lighting system efficacy (lumens/watt)	Hours of operation
Motor efficiency (percent)	Hours of operation
Chiller kW/ton (curve)	Tons of load served
Boiler efficiency (percentage)	Steam produced
Auto (miles per gallon)	Miles driven

It is often necessary in M&V to assign responsibility to the parties involved in an energy efficiency project – and to hold them accountable. Typically, performance factors are the project implementor's responsibility and operational factors are the building owner's responsibility.

Because both performance and operational factors influence impact, an M&V plan must be aligned with the requirements of the energy management contract or program. For this reason, FEMP's *M&V Guidelines* contain a risk and responsibility matrix.[8]

Some examples of performance and operational metric for common systems are listed in Table 2.2.

INSET

> English, like any other language, can be challenging to use in technical situations. The word "performance" has multiple meanings depending on the context. In general, the word "performance" refers to how well something works or operates. However, in a technical sense, performance can refer to a wide range of characteristics, such as speed, efficiency, reliability, and scalability. An M&V professional should involve all stakeholders in a discussion of how the word "performance" will be used throughout the process, in hopes of arriving at a common understanding.

Measuring Performance

Again, language can be tricky. But let us think a little more about performance verification and how it can be instantaneous, short-term, or long-term.

- **Instantaneous:** In this type of performance verification, you take *spot measurements* of the performance of individual components, such as

[8]http://www.eeperformance.org/uploads/8/6/5/0/8650231/femp_m&v_guidelines_v4.0.pdf\(Table~3.1)

HVAC systems, lighting systems, and other energy-consuming devices. The results of these measurements are typically compared to expected or baseline values to determine whether the components are functioning properly.

- **Short-term:** Here, you periodically monitor energy consumption to track performance over a period of weeks or months – whatever is sufficient to evaluate a *performance curve*. This typically entails collecting data from energy meters or building automation systems and comparing energy consumption to expected or baseline values.
- **Long-term:** At this level, you monitor energy consumption and performance over a period of years, typically using an energy use index (EUI) or similar metric. The EUI is a measure of a facility's energy efficiency, calculated by dividing the total energy consumed by the facility by the facility's floor area: kWh per sq. ft. per year.

In the following sections, we will review several programmatic uses of performance indicators that operate at different time scales.

ISO and Other Performance Ideas

The International Standards Organization (ISO) developed a management standard that addresses the development of energy management systems for organizations. Key to these systems is the concept of establishing performance indicators.

Three ISO standards – ISO 50001, ISO 50006:2023, and ISO 50015:2014 – all address energy management but differ in their areas of emphasis.

- *ISO 50001* is the "base" standard. It offers a framework for establishing an energy management system that can help improve an organization's energy performance and reduce its energy consumption and costs. It provides guidance on how to develop and implement an energy management system, formulate energy policies and objectives, and monitor and measure energy performance.
- *ISO 50006:2023* focuses specifically on evaluating energy performance using energy performance indicators (EnPIs) and energy baselines (EnBs). While ISO 50001 guides the development of the framework for the energy management system, ISO 50006:2023 delves into how to gauge and monitor performance.
- *ISO 50015:2014* sets out general principles and guidelines for M&V of the impacts of energy management projects. This standard can work

alongside the other two or independently, and it applies to all types of energy. Its main function is to validate the energy performance results.

Thus, while all three standards touch on energy management, they each have a distinct focus. ISO 50001 provides the basis for an energy management system, ISO 50006:2023 outlines how to evaluate performance within that framework, and ISO 50015:2014 presents standards for conducting M&V. Together, they constitute a comprehensive framework for organizations to manage their energy use and improve their energy performance. The use of EnPIs is a key aspect of this framework, as they provide a standardized approach to measuring and monitoring energy performance.

Building Performance Standards

Buildings are responsible for a significant portion of global energy consumption and greenhouse gas emissions. To address this issue, the energy efficiency industry has developed building performance standards. These standards typically set minimum requirements for building energy performance, such as energy use intensity (EUI) or greenhouse gas emissions per square foot of building space, and may include requirements for building envelope, lighting, heating, ventilation, and air conditioning (HVAC) systems, and other energy-consuming systems. Their fundamental goal is to encourage the adoption of energy-efficient technologies and practices, reduce energy waste, and promote the use of renewable energy sources – ultimately leading to a more sustainable and resilient built environment.
Notable examples of building performance standards include the following:

- In the USA, the Biden administration announced the Federal Building Performance Standard in late 2022. The policy requires federal agencies to cut energy use and electrify equipment and appliances to achieve zero scope 1 emissions in 30% of the building space owned by the federal government (by square footage) no later than 2030.
- The ASHRAE High Performance Building Standard is a voluntary standard, but many building owners and operators use it, as it is recognized by many building codes and rating systems, including LEED and the International Green Construction Code. The ASHRAE standard is a comprehensive set of guidelines and best practices for the design, construction, and operation of high-performance buildings. It is applicable to a range of building types, from commercial and institutional buildings to multi-family residential buildings. Its goal is to provide a road map for building owners, designers, and operators to improve

energy efficiency, indoor air quality, and occupant comfort while minimizing environmental impact. It covers a variety of topics, including building envelope design, HVAC system design and operation, lighting and electrical systems, water conservation, and site sustainability. It also includes requirements for commissioning and ongoing monitoring and maintenance of building systems.[9]
- The Environmental Protection Agency's (EPA) approach to building performance standards is through the ENERGY STAR program. ENERGY STAR is a voluntary program that provides a framework and resources for building owners and operators to improve the energy efficiency of their buildings. The program sets energy performance criteria for different types of buildings, such as offices, schools, hospitals, and retail stores. It also offers tools and resources for measuring, tracking, and improving energy performance. The ENERGY STAR certification is awarded to buildings that meet or exceed the program's energy performance criteria.

Related to performance standards is the concept of benchmarking – a process of comparing the energy performance of a building or facility to a set of established standards or other similar buildings. Building performance standards often set the minimum level of energy efficiency that a building must meet, and benchmarking allows building owners and operators to compare their building's performance to that standard or to other buildings. This comparison can help identify areas where the building is underperforming and can be improved, leading to increased energy efficiency and cost savings. In some cases, benchmarking may be a requirement of building performance standards, as it provides a way to track progress toward meeting energy efficiency goals.

Portfolio Manager is an online tool developed by the EPA to help building owners and managers track and manage energy and water consumption in their buildings. It allows users to enter information about a building, such as its size, age, and energy consumption data, and then benchmark the building's performance against similar buildings in the same region. The tool provides a score between 1 and 100, with a higher score indicating better energy performance. Portfolio Manager is often used to ensure compliance with local energy disclosure ordinances or to track progress toward sustainability goals.

---

[9]https://www.ashrae.org/file%20library/about/government%20affairs/public%20policy%20resources/briefs/approved-ppib---building-energy-benchmarking.pdf

Building Energy Asset Score

The U.S. Department of Energy has developed a standardized tool for calculating Building Energy Asset Scores – numerical ratings of buildings' energy performance based on physical characteristics such as building envelope, HVAC systems, and lighting. The score provides a snapshot of a building's energy efficiency and indicates potential areas for improvement. Because it looks at specific components and systems, the Building Energy Asset Score tool allows for a more granular, detailed analysis of a building's energy performance than building performance standards.

* * *

Combining Energy Asset Scores with Operational Verification (Benchmarking)

It is possible and sometimes informative to combine the function of an energy asset score with operational verification to give a two-dimensional evaluation of an energy system (Figure 2.4).

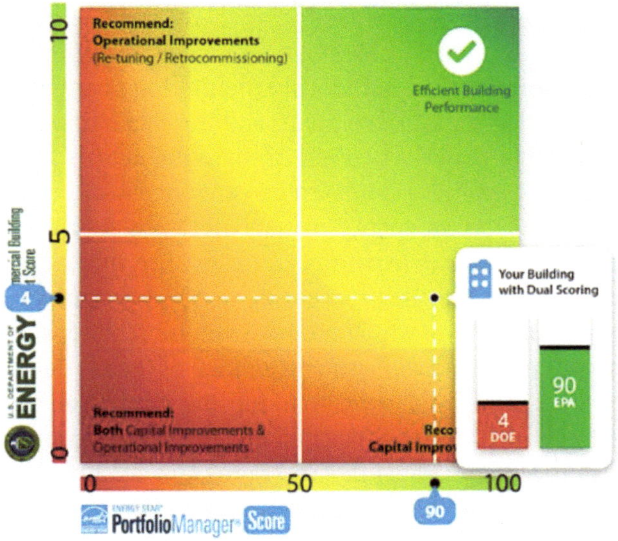

Figure 2.4 Combining Energy Asset Scores with operational benchmarking.

An energy asset score reflects the potential performance of the asset. It is based on performance factors such as building envelope, lighting, and HVAC system characteristics. The score is determined in accordance with a standardized methodology and can be used to compare a building's performance to other facilities with similar operational characteristics. By contrast, operational verification entails measuring the energy consumption of a facility over time to determine if it meets its energy performance targets.

The energy asset score provides information on the installed equipment, while operational verification provides information on the actual energy use of that equipment.

Together, you can use this information to identify areas where energy efficiency improvements can be made as well as to track the effectiveness of energy management projects over time. For example, it is possible for a building constructed with all the most efficient equipment to be operated inefficiently – lights left on, windows left open, controls not working, etc. It is also possible for a building constructed with less efficient equipment to score well on a benchmark metric if it is operated efficiently. The best case is a facility with the most efficient equipment that is also being operated optimally.

SEM

Strategic energy management (SEM) is a comprehensive approach to managing energy use in facilities by continuously monitoring energy consumption, setting energy reduction goals, developing and implementing energy-saving projects, and evaluating the effectiveness of these projects over time.

Performance verification is an essential component of SEM. It is used to ensure that the impacts of energy management projects are in line with expectations and that desired performance levels are maintained. In other words, you can evaluate the effectiveness of an SEM program by using the facility's energy asset score – which provides a comprehensive evaluation of the energy system's performance based on a range of factors such as building type, location, and size – and benchmarking the performance verification results against established targets. You can then use this information to identify areas for improvement, adjust the SEM program's goals, and verify that the desired energy "savings" are being achieved. Performance verification can also point out any operational issues that may be impacting a facility's energy use (Figure 2.5).

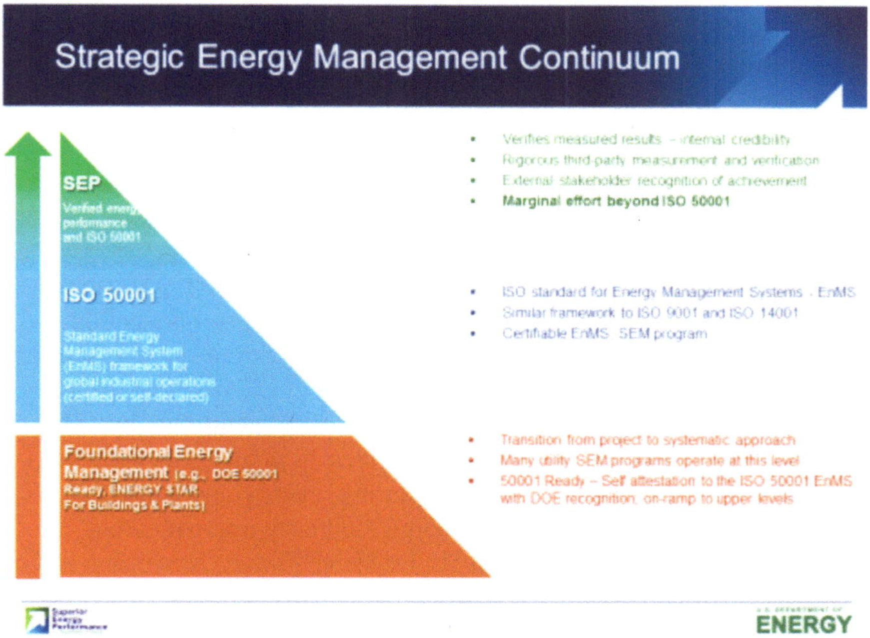

Figure 2.5 Strategic energy management.

Comparing Counterfactual (Impact) to Performance Targets – The Need for Standard Guidance

In many jurisdictions, there is an increasing use of performance targets as a method of promoting energy management. These programs often overlap with programs that use M&V to calculate impact ("savings"). The two activities are closely related. They may appear similar but have different methods for making adjustments in the reporting period.

For example, performance targets such as building performance standards seek to verify that a facility meets a set performance criterion. Programs focusing on impact using the counterfactual method may use different criteria to allow for changes in operation (see non-routine events in Section 4.7).

Without standard guidance, facilities and companies may find it difficult to navigate the different requirements and methodologies of the programs and may end up with conflicting or incompatible results.

Let us hope that regulators and governments will provide clear guidance on how energy management activities can be credited under each regime.

Examples:

Some major cities have implemented ESG requirements or are in the process of developing them:

- New York City: In 2019, New York City passed the Climate Mobilization Act (including Local Law 97), which includes a mandate for large buildings to meet carbon emission limits starting in 2024, with increasingly stricter limits in future years.[10]
- San Francisco: San Francisco has implemented the Existing Commercial Buildings Performance Ordinance, which requires commercial buildings over 50,000 square feet to annually benchmark their energy usage and submit the data to the city.[11]
- Seattle: Seattle has established a goal of carbon neutrality by 2050 and has implemented a number of measures to achieve it, including building energy benchmarking and reporting, energy audits and retro-commissioning, and incentives for energy efficiency improvements.
- Boston: Boston has implemented its Building Energy Reporting and Disclosure Ordinance, which requires buildings over 35,000 square feet to annually report their energy usage and emissions data to the city.
- Chicago: Chicago's Energy Benchmarking Ordinance requires commercial and municipal buildings over 50,000 square feet to annually benchmark and report their energy usage data to the city.

Major cities in Europe and Asia have ESG requirements for buildings. Here are a few examples:

- In the European Union, the Energy Performance of Buildings Directive (EPBD) requires member states to set minimum energy performance standards for buildings and to establish inspection regimes for heating and air conditioning systems.
- The City of London, U.K., has set a target for all new buildings to be net-zero carbon by 2030 and for all existing buildings to achieve net-zero by 2040.
- The City of Paris, France has established the Paris Climate Action Plan, which aims to reduce greenhouse gas emissions by 50% by 2030, with a focus on the building sector.

---

[10] https://www.nyc.gov/assets/nycaccelerator/downloads/pdf/ClimateMobilizationAct_Brief.pdf

[11] https://sfenvironment.org/existing-buildings-energy-performance-ordinance

- In Tokyo, Japan, building owners are required to disclose energy performance data and submit an energy conservation plan every three years.
- Singapore has introduced a Green Mark scheme that assesses the environmental impact of buildings and awards certification based on a range of factors, including energy efficiency, water conservation, and indoor environmental quality.

The Role and Evolution of Protocols/Guidance in Building Codes and Standards

Outcome-based codes are building performance standards that focus on the measurable outcomes of a building's energy and environmental performance, rather than prescriptive design requirements. Rather than specifying particular technologies or design features, outcome-based codes set performance targets that buildings must meet to comply with the standard.

The key principle behind outcome-based codes is that building owners and designers are given greater flexibility to achieve energy and environmental performance targets in the way that works best for their building and context. This allows for greater innovation and creativity in building design and can help drive the market for new energy-efficient technologies and practices.

One example of an outcome-based code is the Zero Code, developed by the New Building Institute (NBI) in the United States. The Zero Code sets a performance target of net-zero energy use for new buildings, as well as targets for water conservation, indoor air quality, and other environmental and health outcomes.

Outcome-based codes are still relatively new and are evolving rapidly. As more cities and jurisdictions adopt outcome-based codes, it is likely that they will become an increasingly important tool for driving innovation and sustainability in building design and construction.

M&V professionals may be responsible for measuring and verifying the outcomes required by outcome-based codes.

Special Case: Simulated Baseline for New Construction

One exception to the pre/post (counterfactual) structure of M&V is when M&V is applied to a new construction project. In such projects, M&V ensures that the energy-efficient features of the facility are designed, installed, and functioning correctly. This is done using modeling and simulation techniques, as well as on-site testing and inspection during construction.

The M&V process for new construction typically follows these steps:

- **Establishing a baseline:** The energy use of the facility is estimated based on its design and expected occupancy.
- **Monitoring during construction:** The performance of energy-efficient features is tested during construction to verify that they meet design specifications.
- **Testing and commissioning:** The facility and its components undergo testing and commissioning to ensure that all energy-efficient features are functioning properly.
- **Post-construction monitoring:** The facility's energy use is monitored and compared to the baseline to verify that the energy-efficient features are performing as expected.
- **Ongoing monitoring:** The facility's energy use is monitored on an ongoing basis to identify any issues or opportunities.

In general, M&V for new construction projects is more complex and data-intensive than in retrofit projects, due to the need to develop a baseline "from scratch."

The Counterfactual Method in Other Disciplines

As shown, the counterfactual method is crucial to the work of M&V. It is not unique to the profession, however. Numerous other disciplines also apply it to evaluate the impacts of interventions.

- **Program evaluation:** The counterfactual method is used to assess the impact of social programs and interventions, such as education programs, health interventions, and job training programs.
- **Epidemiology:** Epidemiologists use the counterfactual method to assess the effectiveness of medical treatments and interventions.
- **Policy analysis:** The counterfactual method can be applied to evaluate the effectiveness of government policies and programs, such as education or healthcare initiatives.
- **Econometrics:** Here, the counterfactual method can reveal the causal relationships between different economic variables, such as the impact of changes in interest rates on consumer spending, or the effect of inflation on unemployment rates.
- **Market research:** Market researchers use the counterfactual method to measure the effectiveness of advertising and marketing campaigns and assess consumer behavior.

While the terminology used in each of these disciplines may be slightly different, the underlying concepts and techniques are similar. They all involve comparing outcomes before and after an intervention or treatment, and then using a counterfactual to estimate *what would have happened* in the absence of the intervention.

2.1.10 Uncertainty

Uncertainty is a reality. It is not a question of whether it exists – it is only a question of whether to deal with it explicitly or ignore it (aka denial). In M&V, we can use three distinct frameworks to consider uncertainty. The M&V professional may find this one of the most challenging and interesting parts of their job. The ability to command and communicate uncertainty concepts will greatly enhance your chances of successful settlement.

Uncertainty in Science

In the sciences, uncertainty refers to the limitations and potential errors inherent in any scientific measurement, observation, or experiment. These uncertainties can arise due to a variety of factors, such as the limitations of measuring instruments, natural variations in the phenomena being studied, or errors in experimental design.

In science, uncertainty is typically quantified using statistical analysis techniques. These techniques help scientists evaluate the reliability and significance of their results and identify potential sources of error or bias.

When scientists present their findings, the best practice calls for reporting the degree of uncertainty. This allows other researchers to evaluate the reliability and relevance of the findings and potentially build upon them in future research.

Uncertainty in Engineering

Engineers deal with uncertainty differently from scientists. Engineers often incorporate "factors of safety" into their designs to account for uncertainties in loads, material properties, and other things that can affect performance. Factors of safety provide a margin of error so that the system will operate safely even if actual conditions deviate from the design assumptions.

Uncertainty in Risk Management

Risk managers use statistical and probabilistic methods to quantify uncertainty and develop models that can help predict the likelihood and potential impact of different scenarios. By understanding the sources of uncertainty and the potential consequences of different outcomes, they can develop strategies to manage and minimize risk.

M&V professionals may also use risk analysis drawing on several different techniques:

- One option is to incorporate *contingency factors* – or "risk buffers" – when estimating a project's energy impacts. Like factors of safety, these buffers safeguard against potential variations in outcomes due to uncertainty.
- *Probabilistic modeling* quantifies the probability of various outcomes based on different assumptions and inputs, allowing for more informed decision-making.
- A *sensitivity analysis* examines how and to what extent changes in inputs or assumptions affect the overall energy impact estimate. This way, we can identify which inputs or assumptions have the greatest effect on the project's outcome and where additional data or analysis may be needed to reduce uncertainty.
- Other risk management techniques such as *Monte Carlo simulation* and *stochastic modeling* also assist M&V professionals in accounting for uncertainty and risk.

We will discuss these techniques in greater detail in Chapter 4.

Identify, Quantify, and Manage

As an M&V professional, you cross disciplinary lines, and when it comes to your view of uncertainty and risk, the right balance will be different depending on the context. For an ESPC project, it might be roughly 25% science, 25% practical engineering, and 50% risk-sharing. As such, it is your responsibility to convey the significance of uncertainty to engineers and other, perhaps non-technical, stakeholders who may view the concept of uncertainty through a different lens. Here are some tips on how to communicate uncertainty effectively:

- **Define terms:** Start by defining technical terms and concepts in a way that is easy for all parties to understand. Use plain language and avoid jargon.
- **Explain the importance:** Help stakeholders understand why uncertainty matters to the project and what impact it could have on the outcomes. Use real-world examples to illustrate the point.
- **Acknowledge uncertainty:** Listen to the concerns of other stakeholders and address them directly. Acknowledge any uncertainty and

work together to develop strategies to adopt a common language and framework.

Consider this example of how effective communication about uncertainty might play out in an M&V context. Imagine that you are an M&V professional working on a project to upgrade a facility's chiller. You are meeting with the lead engineer and the risk manager.

You: You know, we have got to think about uncertainty here. Take the cooling load, for example. It can fluctuate *a lot* – depending on the day's weather, how many people are in the building, and even what machines are running.
Engineer: I am not too concerned. We can just oversize the chiller to handle any variations in load.
You: Oversizing the chiller will increase the initial cost *and* operating costs. Remember, we are trying to cut down on costs. Why don't we try a different approach? Let us crunch some numbers, do a little statistical analysis, and figure out the load we are likely to deal with. We can then get a chiller that fits the bill without going overboard.
Risk manager: What if the chiller has an unexpected breakdown or some downtime? That is going to impact performance and our energy savings.
You: We will talk more about the word "savings" later – otherwise, that is a solid point. We should include a risk analysis in the M&V plan. We can also keep a close eye on the chiller's performance to detect any issues early on.

In this example, you point out the need to consider uncertainty in the chiller project, and the lead engineer proposes a solution based on a factor of safety approach. But if the engineer selects a chiller with a higher cooling capacity than required for the expected load, this overdesign could result in the chiller running inefficiently at partial loads, reducing the energy management project's overall impact. You suggest instead using statistical analysis to estimate the load with a certain degree of confidence. When the risk manager brings up the possibility of unexpected failures or downtime, you suggest including a risk analysis in the M&V plan and continuously monitoring the chiller's performance.

At some point early in the process, you may want to conduct a performance test on the chiller to measure the actual cooling capacity and efficiency at various loads. Based on this data, you can build a model to predict system performance under different conditions and quantify the uncertainty associated with these predictions. This way, your calculations can accurately reflect the true performance of the system and account for any uncertainty or variability that may arise in actual operation.

However, simply by communicating the importance of understanding the true performance of the system and quantifying uncertainty, you have already helped ensure that the design and operation of the system are optimized for efficiency, not a factor of safety.

2.1.11 Cost

Managing the costs of M&V is a key aspect of your job as an M&V professional. M&V can be costly, especially for large or complex projects. An important skill to develop is to be able to determine the "right" amount of M&V to apply for any given project. This requires professional *judgment* because, in M&V, we can never arrive at a "perfect" answer – some degree of uncertainty is unavoidable. More M&V can reduce that uncertainty, but collecting and analyzing data can get very expensive very quickly. For this reason, when considering costs, one of your first steps should be to take stock of the needs of the primary stakeholders. What financial resources are they willing to spend on M&V? What is their appetite for risk – are they willing to tolerate higher levels of uncertainty? The answers to these questions should inform your judgment about the optimal "amount" of M&V.

As a general practice – but especially if you find that you must minimize costs on an M&V project – it can be a good idea to take some of the following measures:

- **Prioritize which data to collect:** In M&V, not all data is equally important; so it is important to prioritize which data to collect based on its importance for achieving the project's goals. This way, resources can be focused on the most critical data, rather than being spread too thin.
- **Choose the right tools for the job:** As an M&V professional, you need to use a variety of tools and technologies to collect and analyze data. Using the "right tools for the job" streamlines data collection and analysis and reduces the time and effort required for M&V.

These general strategies can help you take steps to manage the costs of M&V projects while still achieving project goals.

Cost management is essential to the work of M&V, but it is not enough. As an M&V professional, you must also be able to explain costs and convey the value of M&V to stakeholders. This might mean illustrating the potential for cost savings or better resource efficiency through M&V and outlining how these outcomes align with stakeholder needs and objectives. In this way, you can frame the expense and efforts invested in M&V as a strategic decision rather than a simple operational cost.

* * *

Note: As of now, M&V is still reserved for larger projects. Except for simple performance measurements, it remains too expensive for most small and medium markets.

* * *

Two Predictions – Targets/Outcomes and What Would Have Happened

By now, I hope we have convinced you that M&V is based on either 1) performance verification or 2) long-term impacts as quantified by comparing actual energy use to a hypothetical (counterfactual), aka the adjusted baseline.

But that is only part of the story. What about the effectiveness of the energy management activity as compared to the predictions that led to the investment?

Energy management is like science, in that energy engineers develop a hypothesis about how investments can be made to impact the energy consumption of a facility. The counterfactual method reveals the amount of that impact. But along with deploying the counterfactual, it is a good practice to report whether the prediction was accurate. This adds a third "line" to the classic M&V graph: the adjusted baseline, the actual performance, and the predicted performance. Can we call this the "future factual"? Probably not. There is another energy management technique known as monitoring and targeting[12] that addresses the concept of maintaining a performance target and maintaining feedback through monitoring.

The M&V professional may well run into M&T projects and find much of the terminology and goals very similar. We are all in this together (Figure 2.6).

Value of Information, Judgment, and Risk in the Real World

The term "value of information" (VOI) refers to the utility of certain information or data in decision-making. For example, obtaining accurate measurements for M&V can reduce uncertainty and help identify and manage

[12]https://en.wikipedia.org/wiki/Energy_monitoring_and_targeting

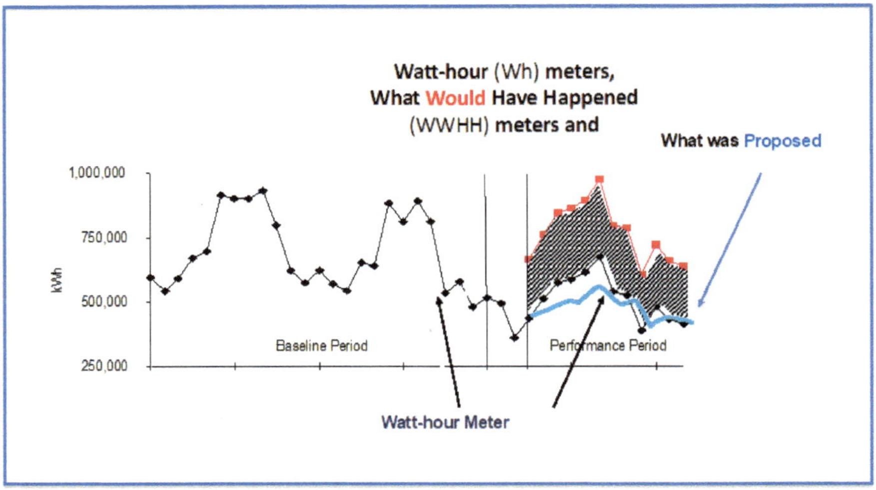

Figure 2.6 Adding a line of proposed performance to the classic M&V graph.

risks. But VOI can also be considered in a probabilistic or statistical sense. In this case, it is often discussed in the context of Bayesian decision theory. Here, VOI is defined as the expected improvement in decision quality that results from having that information. In other words, how much more accurately can we make decisions, given this additional piece of information?

Because VOI helps us identify the most significant inputs or parameters in a system – as well as the areas where the need for additional data is most critical – it plays an important part in the development of models for M&V. A VOI analysis quantifies the expected reduction in uncertainty or the expected improvement in decision-making that would result from obtaining additional information. This can be used to identify which areas of the model would benefit most from additional data or more detailed measurements. In this way, VOI analysis helps us set priorities and determine how to allocate resources to obtain the most information at the least cost. Similarly, a *sensitivity analysis* is a method for evaluating how the output of a model changes as a function of the inputs or parameters. The results of this analysis can be used to identify which inputs or parameters have the greatest impact on the model's output.

Risk Framework for Benefits

There is generally an inverse correlation between uncertainty and value, meaning that as uncertainty decreases, the value of a decision or outcome increases.

When faced with high uncertainty, we find it difficult to predict the outcomes of different options – and, by extension, make informed decisions. This is because there is a greater risk that the decision will not lead to the desired outcome. As a result, the value of the decision or outcome is often lower. Conversely, as uncertainty is reduced, our decision-making improves, and the value of a decision or outcome increases because the potential risks and negative consequences of a decision are better understood and can be more effectively managed.

There are several ways to reduce uncertainty, such as by collecting additional data, improving the quality of data, developing more accurate models, or implementing better risk management strategies.

That said, it is impossible to *completely* eliminate uncertainty. And we should also acknowledge that the relationship between uncertainty and value is not linear. There are *diminishing returns*: the value of reducing uncertainty decreases as uncertainty decreases further.

Imagine that a company is considering investing in new technology for their manufacturing process. Unfortunately, the company is uncertain how well the technology will work, how much it will cost to implement, and what the potential return on investment will be. The higher the uncertainty, the less confident the company will be in its decision to invest in technology. But if the company takes steps to reduce uncertainty – by conducting additional research, developing more accurate prediction models, etc. – it will be more confident in its decision.

As an M&V professional, you should nearly always try to spend the least amount possible. If there is no perceived value for additional expenditures on M&V, it is not prudent to use more resources.

Uncertainty Reduction and Cost: The Tradeoffs

One method of quantifying the value of reducing uncertainty is decision analysis, which provides a framework for making decisions under uncertainty. We can use decision analysis to estimate the value of perfect information (VPI), the maximum amount that we would be willing to pay for additional information that would eliminate all uncertainty about a decision.

The VPI is based on the difference between the expected value of a decision with perfect information and the expected value of the decision with the information that is currently available. To use a financial metaphor, VPI can be thought of as the value of a bond or other financial instrument that pays a fixed amount if a particular event occurs. For example, imagine that you are considering buying a bond that pays $100 if a certain company goes bankrupt. The expected value of the bond is the probability of the company going bankrupt multiplied by the payout if it goes bankrupt, discounted by the cost of the bond. The VPI is the difference between the expected value of the bond with perfect information about the company's bankruptcy and the expected value of the bond with the information that is currently available.

One way to quantify the value of reducing uncertainty is to estimate the cost of risk. The cost of risk is the cost associated with the potential negative consequences of a decision, such as the cost of project delays, litigation, or damage to reputation. By estimating the cost of risk, we can compare the cost of reducing uncertainty through M&V activities to the potential cost savings associated with avoiding negative consequences.

2.1.12 Concluding this Section

M&V methods come from established fields of study such as engineering, physics, statistics, and economics. For example, engineering principles inform the design and optimization of energy systems, the laws of physics elucidate the behavior of energy in facilities and systems, statistical analysis helps quantify the impact of energy management activities, and approaches from economics ensure the financial viability and cost-effectiveness of energy management projects.

To some extent, M&V has only borrowed concepts and applied new terminology. This has been beneficial in areas where energy projects require specialized language. However, it isolates us from potential contributors outside our profession. For this reason, when working within any team, you should first determine if those involved already understand the fundamental concepts, using different terminology.

Is the counterfactual method science or philosophy? Or the science of philosophy? It is appearing in a range of areas these days, including machine learning and artificial intelligence – and human consciousness. As Judea Pearl and Dana Mackenzie write in *The Book of Why: The New Science of Cause and Effect*:

> Counterfactuals are an essential part of how humans learn about the world and how our actions affect it. While we can never walk down both the paths that diverge in a wood, in a great many cases we can know, with some degree of confidence, what lies down each.

While the above is true, some people distrust methods that allow for a hypothetical adjusted baseline. They only trust what they can see in front of them in real time. In these situations, it is best to learn that before you begin to develop the M&V plan. Do not engage in an activity that will only lead to contention. Perhaps in these situations, monitoring and targeting will be a better solution.

The *fundamental role of the M&V professional* is to facilitate agreement among all stakeholders that the reported quantity fairly and accurately represents the impact of the relevant energy management activities. To borrow the words of Pearl and Mackenzie, we must ensure that stakeholders "know, with some degree of confidence, what lies down each" diverging path – both the path that we have taken and the one that we have not. Successful settlement is the end in mind, and well-designed M&V is the best path to successful settlement.

In the next section of this chapter, we will explore the variety of contexts in which M&V occurs and the importance of understanding the specific requirements of each.

Exercises and Thought Experiments

During your next M&V project, ask the stakeholders to define the following terms: *savings*, *performance*, *operation*, and *counterfactual*.

* * *

If the owner shuts down part of a facility, have "savings" been achieved? Why or why not?

* * *

Imagine that you are an M&V professional working with a facility owner who wants to implement an energy efficiency project. The owner has a limited budget and is considering between two M&V methods: performance measurements and counterfactual M&V.

The benefits of performance measurements include the following:

- **Cost-effective:** Performance measurements are generally less expensive because they require fewer resources.
- **Real-time monitoring:** Performance measurements can provide real-time data, which allows for quick identification of potential issues or opportunities for improvement.

- **Transparency:** Performance measurements are easy to understand and communicate, which can help to build trust with facility users and stakeholders.

However, relying solely on performance measurements also carries some risks:

- **Limited scope:** Performance measurements might not capture all the factors that affect energy use in a facility, such as changes in weather or occupancy patterns.
- **Greater uncertainty:** Performance measurements might not provide a clear indication of the impact of an energy efficiency project over time, which can make it difficult to determine the return on investment.

On the other hand, counterfactual M&V offers the following benefits:

- **More accurate assessment over time:** Counterfactual M&V can provide a more accurate estimate of the impacts of an energy management project over time, which can help ensure that the project meets its goals.
- **Comprehensive:** Counterfactual M&V accounts for a wide range of factors that can affect energy use in a building, giving a more comprehensive view of the project's impact.
- **Independent:** Counterfactual M&V can be conducted by an independent third party, which can help build trust and stakeholders.

However, counterfactual M&V also presents some risks:

- **Costly:** Counterfactual M&V is nearly always more expensive than performance measurements because it requires more resources.
- **Time-consuming:** Counterfactual M&V requires collecting and analyzing data throughout the reporting period, delaying the reporting of results/success of the project.
- **Complex:** Counterfactual M&V is more complex, making it more difficult to communicate and explain the results to stakeholders.

Ultimately, the decision will depend on the specific needs and goals of the facility owner and other stakeholders, as well as the budget and resources available.

2.2 Contexts

The scientific method and the M&V methods derived from it are general and global. The fundamentals of M&V, as a subset of science and engineering, are also global constructs. They know no borders. They apply generally.

However, when an M&V professional applies these concepts in the course of a specific project, the resulting M&V activities may vary widely depending on the context. A study of the impacts of a new technology in a California utility program evaluation will likely call for a very different approach than a lighting retrofit in a performance contract in Sri Lanka. They are both M&V – but they are also M&V *performed within a specific context.*

The context of an energy project can include factors such as the location of the project, the type of energy source being used, the regulatory environment, and the availability of financing and resources. Understanding the context of a project is essential. It allows you as an M&V professional to identify potential challenges and opportunities and tailor the project to the context's specific needs and requirements.

The first thing you must ask is "What is the context in which the M&V will occur?" For example:

- Is the project part of a performance contract? A utility program? A government program?
- Are there governing rules to follow, or are the parties free to establish any contract they want?
- Are there pre-established methods for determining "savings"?

The answers to these questions will be the foundation on which to plan exactly how to execute M&V for the project.

2.2.1 Domains

As an M&V professional, you are an engineer whose work goes well beyond engineering. It is possible to think of your work as spanning four domains, or areas of practice:

- **Regulatory:** In the regulatory domain, you must understand, navigate, and comply with local, national, and international laws and regulations – whether they relate to energy efficiency, carbon emissions, environmental protections, or industry-specific requirements. You will need to ensure that projects are compliant with these standards, and you might even participate in developing regulatory reports or audits. Of course, all this means that you will have to keep up to date with any regulatory changes that could impact your projects.
- **Contractual/programmatic:** Your responsibilities are largely governed by the terms of contracts and program guidelines, which dictate specific methods, timelines, and performance metrics for M&V tasks. You

must ensure strict adherence to these terms, provide clarification to all involved parties, and handle deviations from the contract or guidelines. Additionally, you play a crucial role in avoiding and, if necessary, managing disputes.

- **Economic:** This part of your work centers on the economic aspects of energy projects. The most important financial task will likely be to arrive at the amount of cost savings for the project. This will involve working with others to establish the value of the energy impact, as determined by using a rate/tariff structure, published avoided cost values, or an agreed-upon valuation method. But you must also be cognizant of the cost of M&V and be able to assist all parties in understanding cost and risk tradeoffs.
- **Physical (engineering):** This is perhaps the most obvious aspect of your work as an M&V professional. It involves the actual measurement and verification of equipment performance, energy usage, and modeling. This often entails conducting site visits, installing and managing measurement equipment, analyzing data from energy systems, and creating baseline models.

As an M&V professional, you should understand all the relevant details, *across all domains*, of M&V's role in a specific project. Is the M&V part of the settlement for a contract where risk is shared between parties? Is the M&V part of a utility program with specific, predetermined requirements? Will other government agencies review the results and use them for reporting? All these questions should be answered before the M&V process begins.

Table 2.3 Four domains that define the context of M&V.

Common considerations in each domain	
Regulatory • Government policies • Market transformation • Carbon or climate goals • Grid impacts	**Contractual/programmatic** • Risks/responsibilities of stakeholders • Dispute resolution • Force majeure • Contracts – government, private party, etc. • Shared savings • Guaranteed savings
Economic • Rates/tariffs • Credits/certificates • Utility incentives • Financing incentives • Tax incentives	**Physical** • Facility audit data • Performance data • Usage data • Weather data

We will discuss each domain in greater detail below.

Regulatory

The regulatory domain is a critical component of the overall energy management landscape. Governments develop and implement regulatory and policy frameworks to promote energy efficiency, renewable energy, and other clean energy initiatives. Some examples of these policies and initiatives include:

- **Renewable portfolio standards (RPS):** These policies require utilities to generate a certain percentage of their electricity from renewable energy sources.
- **Energy efficiency standards:** These policies establish minimum energy efficiency requirements for appliances, buildings, and other energy-consuming products.
- **Tax credits and incentives:** Many governments offer tax credits and other incentives to encourage the development and adoption of clean energy technologies, such as solar panels, wind turbines, and energy-efficient appliances.
- **Energy efficiency programs:** Government agencies often offer energy efficiency programs and services to help consumers and businesses reduce their energy consumption and lower their energy bills.
- **Environmental regulations:** Regulations such as the Clean Air Act and the Clean Water Act (in the United States) are designed to protect public health and the environment by reducing air and water pollution from energy production and consumption.
- **Permitting and licensing:** Governments play a critical role in permitting and licensing energy projects, including both fuel-based and renewable energy projects such as solar and wind farms.

As this list suggests, the regulatory domain of energy projects is complex and multifaceted, and it requires careful attention. In most contexts, energy efficiency standards and building codes mandate minimum levels of energy efficiency for new buildings or major renovations. These standards specify requirements for insulation, lighting, HVAC systems, and other building components. Also relevant are regulations related to energy consumption and reporting, such as requirements for companies to report their energy usage and greenhouse gas emissions to government agencies.

As an M&V professional, you are responsible for understanding and complying with relevant regulations and standards governing energy efficiency and M&V. To this end, you might often work with energy consultants, who provide expertise on regulations and guidelines and advise on compliance strategies. Occasionally, you might deal directly with regulatory agencies to ensure project compliance with regulatory requirements and obtain necessary approvals.

For utility programs specifically, you may need to work with program managers and engineers to ensure that the M&V plan is aligned with program requirements and that the impact calculations are consistent with the program's technical and financial parameters. You may also need to coordinate with the utility program personnel to gather data, review utility bills, and ensure that the M&V process is properly documented and reported to the program administrators.

Contracts and Programs

Energy management projects can be delivered either under contractual arrangements or through programs. An *energy savings performance contract* (ESPC) is a contractual arrangement between an energy services company (ESCO) and a customer, typically a building owner or a facility manager.

In this type of arrangement, the ESCO provides energy management services and solutions in exchange for a fee. The scope of services in an energy management contract can be customized to meet the specific needs of the customer, and it can include various services, such as energy procurement, demand response, energy data analytics, and energy project development. Often, the ESCO implements one or more energy-saving measures, such as lighting upgrades or HVAC system improvements, and the customer "pays back" the ESCO a portion of the energy savings over a specified period of time. Various financial structures exist to share the risk of implementing the energy-saving measures. In Chapter 4, we briefly describe the most common forms.

Many utilities, private companies, and governments offer *energy management programs* to their customers or citizens to help them reduce their energy consumption and costs. These programs promote energy management that can consist of various elements, from energy audits and benchmarking to education and outreach, financial incentives, and regulatory policies. They provide rebates, incentives, or financing to participants who implement energy efficiency measures, such as adding insulation or upgrading to more efficient appliances. The goal of such programs is to reduce energy consumption and improve energy efficiency across a region, sector, or group of buildings. Types of programs include:

- **Utility programs:** Customers may have the opportunity to participate in energy efficiency programs implemented by utilities, sometimes implemented in conjunction with ESCOs or other energy providers.
- **Public sector programs:** Many state and local governments have established energy efficiency programs for public sector entities, such as schools and government buildings, offering them financing or technical assistance.
- **Private sector programs:** Private sector companies may also offer energy efficiency programs to their customers or tenants. For example, a commercial building owner may financially incentivize tenants to reduce their energy costs and improve the building's energy performance.

The financing and incentives provided by these programs help make the upfront cost of energy efficiency measures more manageable for participants in service of a larger goal.

While both approaches – contracts and programs – aim to improve energy efficiency, the key difference is that an energy management contract is an agreement between a customer and an ESCO, while an energy management program is typically initiated by a government agency or utility company and involves multiple stakeholders. Programs are often designed to address specific energy efficiency goals, such as reducing greenhouse gas emissions or improving grid reliability, while energy management contracts are tailored to meet a particular customer's needs.

Contracts

Energy savings projects often involve contracts between various parties, including the energy service provider (ESP or ESCO), the customer, and other potential stakeholders, such as financing institutions. ESCOs – which have been around since the 1970s[13] – can provide a range of services, from single measure installations (lighting retrofits) to comprehensive energy projects for complex facilities. Contracts may specify the terms and conditions of the project, including the scope of work, the expected energy savings, the payment structure, and other important details. The legal implications of these contracts can be significant, as they can impact the liability and responsibility of each party involved in the project. For example, the contract may specify who is responsible for the installation and maintenance of energy efficiency equipment, who bears the risk of equipment failure or underperformance,

[13] https://en.wikipedia.org/wiki/Energy_service_company

and who is entitled to any financial incentives or penalties related to energy savings performance.

As an M&V professional, you may need to work with attorneys, contract managers, and procurement officers to ensure that the M&V plan and activities comply with legal and contractual requirements. You may also work with project management personnel to develop risk mitigation strategies and ensure that the M&V plan is consistent with the project's financial and technical goals. In some cases, you might even assist in drafting or reviewing the M&V-related sections of contracts, including performance specifications, payment terms, and dispute resolution procedures.

At minimum, you should understand the terms of the contract, with specific focus on areas that overlap with the technical performance of the project. Some clauses to know include:

- **Replacement:** This type of clause defines the responsibilities and procedures for replacing equipment or components of the energy management system. The clause may specify conditions under which replacements are required, who bears the cost of replacement, and the timeline for executing the replacement.
- **Controls:** Control clauses outline the requirements for the control systems installed as part of the energy management project. They may include specifications regarding the type of controls, their functionality, set points, and performance criteria. The purpose of a control clause is to ensure that the control systems are properly installed and maintained to achieve energy efficiency goals.
- **Tune-ups:** Tune-up clauses define the procedures and frequency of tune-ups of the energy management system. Tune-ups involve adjusting, calibrating, and optimizing the system so that it continues to operate at or near peak performance. The clause may outline who is responsible for conducting the tune-ups, the schedule for tune-up activities, and any associated costs.
- **Maintenance:** This type of clause outlines the maintenance requirements for the energy management system. It specifies the responsibilities of the parties involved in terms of routine inspections, cleaning, and servicing of equipment. Maintenance clauses may also cover the replacement of consumables, such as filters or lubricants, and the frequency of maintenance activities.
- **Labor:** Labor clauses detail the labor requirements and responsibilities associated with the energy management project. They may specify the number and qualifications of personnel required to perform the M&V

activities, the duration of their involvement, and any relevant labor rates or cost provisions.

- **Non-energy benefits:** These clauses address the non-energy benefits that may result from the implementation of the energy management project. Non-energy benefits can include improved occupant comfort, indoor air quality, productivity gains, and reduced maintenance costs. The contract may lay out how to measure and verify these benefits and any related obligations or compensation.
- **Non-routine events detection (static factor logs):** These clauses pertain to the detection and handling of non-routine events. Static factor logs are typically used to identify deviations from expected system behavior. The contract may outline the procedures for detecting and addressing these events, including the parties responsible for monitoring and responding to alerts or alarms.
- **Non-routine adjustment methods:** A non-routine adjustment method clause defines the methods and protocols for any non-routine adjustments necessary during an M&V. It may specify the approval process, documentation requirements, and responsibilities for implementing these adjustments.

Particularly pertinent to "M&V' was the development of the FEMP program in the early 1990s – the first large "Super ESPC." This new program introduced standardization in procurement and M&V, as we will see in Chapter 3.

Countries with state sponsored ESCO programs currently include:

- United States: The U.S. government has several programs that promote energy efficiency and renewable energy in buildings, including the Federal Energy Management Program (FEMP) and the Energy Savings Performance Contracting (ESPC) program.
- Canada: Canada's Energy Services Acquisition Program (ESAP) provides federal government departments and agencies with access to energy efficiency services and solutions.
- Saudi Arabia: The National Energy Services Company (Tarshid) is a Saudi Arabian government entity responsible for implementing energy efficiency and conservation initiatives. Tarshid plays a significant role in driving energy efficiency measures in various sectors of the economy.
- United Kingdom: The U.K.'s Re:fit program is a procurement initiative that helps public sector organizations reduce energy consumption and carbon emissions.

- Germany: Germany has several government programs that promote energy efficiency in buildings, including the Energy Performance Contracting (EPC) program and the KfW Energy Efficiency Program.
- Japan: The Japanese government's Top Runner program works to improve energy efficiency in buildings and appliances by setting efficiency standards based on the most efficient products and technologies available.
- South Korea: The South Korean government's ESCO project promotes energy efficiency in buildings and industry by providing financial incentives and technical support to ESCOs.
- United Arab Emirates: The UAE's Super ESCO program attempts to accelerate energy efficiency investments and reduce energy consumption in the country.
- Malaysia: Malaysia's Super ESCO program promotes energy efficiency and reduces carbon emissions in the country.
- Vietnam: The Vietnamese government's Super ESCO program likewise promotes energy efficiency and reduces energy consumption in the country.

These are just a few examples. Many other countries have established similar programs.[14]

Programs

Whereas contracts for energy services make up a large percentage of energy management activity, programs represent a large and growing portion of M&V.

Here are some examples of utility and tax incentive programs:

- **Energy efficiency rebate programs:** Rebate programs incentivize customers to upgrade to more energy-efficient equipment or appliances. For example, a utility may offer a rebate for installing a high-efficiency air conditioning system or upgrading to LED lighting.
- **Time-of-use rates:** Time-of-use (TOU) rates incentivize customers to shift their energy usage to off-peak hours, when energy costs are lower.
- **Demand response programs:** Demand response programs incentivize customers to reduce their energy usage during periods of high demand.

[14]Global ESCO Network (GEN): The Global ESCO Network is an initiative by the United Nations Environment Programme (UNEP) that aims to facilitate information exchange and collaboration among ESCOs, policymakers, financial institutions, and other stakeholders.

Customers may receive financial incentives for reducing energy usage during these periods.

- **On-bill financing:** On-bill financing programs allow customers to finance energy-efficient upgrades through their utility bill. This can make it easier and more affordable for customers to make energy-efficient upgrades.
- **Property tax incentives:** Some local governments offer property tax incentives for energy-efficient buildings. For example, a building owner may receive a property tax reduction for installing energy-efficient windows or insulation.
- **Federal tax incentives:** The U.S. federal government offers tax incentives for energy-efficient upgrades, such as the Commercial Building Tax Deduction and the Residential Renewable Energy Tax Credit. These incentives can help reduce the cost of energy-efficient upgrades for building owners and homeowners.

Many types of programs have evolved, as utilities, governments, and private organizations have sought to meet the needs and challenges of the energy transition. And while all these types of programs can involve some M&V, it is most likely – at least in the U.S.A. – that an M&V professional will play a role in a custom or pay-for-performance program.

Custom programs give customers incentives for implementing energy efficiency measures not covered by standard rebate programs. These measures are often tailored to the specific needs of the customer and can include things like advanced lighting controls, building automation systems, or customized HVAC solutions. The customer and utility work together to identify energy-saving opportunities and determine the appropriate incentive amount. The incentive is typically based on the estimated energy impact of the project.

Pay-for-performance programs incentivize customers based on the actual impacts of energy management measures. These programs tend to require the installation of advanced metering equipment that allows for real-time monitoring of energy usage. Customers are then paid based on the actual energy impacts achieved, rather than a predetermined incentive amount. Pay-for-performance programs are more complex than standard rebate programs, but they offer the potential for greater energy management impacts and more accurate measurement. They also incentivize customers to implement ongoing energy-conversation measures and continuously monitor their energy usage.

Both custom and pay-for-performance programs are offered by utilities as part of their energy efficiency portfolio. Not only can they provide significant financial incentives for customers to implement energy-conservation measures, but they also help utilities reduce overall energy demand and improve grid stability.

Programs use M&V as a general ruleset to assess the impacts related to their specific focus. For example, in California, the strategic energy management program has its own M&V manual.[15]

As an M&V professional, you must be aware of a program's rules and guidelines and how they constrain the way an M&V will be conducted. This is true regardless of the program type.

Flex Markets

Up until recently, most energy management programs focused on reducing end-use energy consumption. Market transformation introduced new and efficient technologies, and attention to operational characteristics optimized the utilization of existing assets. But in the past decade, the delivery of energy has evolved and created new requirements to maintain an effective grid/distribution system. In many jurisdictions, it has become equally important to manage *when* energy is consumed. Shifting loads, rather than reducing them, can also be valuable. Programs to reward load shifting will require new and different approaches to baselining and "M&V."[16]

Uncertainty, Risk, and Error – and How They Differ for Contracts and Programs

Uncertainty, risk, and error are important considerations in both ESPC and programs. However, the approaches for addressing uncertainty, risk, and error often differ from one context to another.

In ESPC, contractually stipulated forms of risk allocation and performance guarantees attempt to minimize risk and uncertainty. For example, a contract might specify performance metrics that must be met in order for the contractor to be paid, or may require the contractor to bear some of the financial risks associated with the project. The contract may also include provisions for M&V.

---

[15]https://pda.energydataweb.com/api/view/2648/CA_SEM_MV_Guide_v3.02.pdf

[16]"Flex markets" typically refer to markets where flexibility in energy supply and demand is bought and sold. These markets enable the trading of various forms of flexibility, such as demand response, energy storage, and distributed energy resources, to help balance the electricity grid and optimize its operation.

Programs tend to address uncertainty and risk through statistical analysis and data validation. Evaluators often use statistical models to estimate the impact of the program on energy consumption or other metrics. They may also conduct sensitivity analyses to assess the impact of different assumptions or inputs. Data validation ensures that the data used in the analysis is accurate and reliable.

Both ESPCs and the evaluation of utility programs must confront the possibility of error. In ESPCs, error can stem from inaccurate energy impact estimates or unforeseen changes in building occupancy or operations. In programmatic contexts, error may be due to incomplete or inaccurate data or from unwarranted assumptions or simplifications made during the analysis.

Besides these differences, though, both contractual and programmatic approaches to error usually incorporate sensitivity analyses and uncertainty assessments to help identify potential sources of error and quantify their impact on the results. In other words, while some of the specific techniques used to address uncertainty, risk, and error differ between contracts and programs, both share a common goal of ensuring that the impacts of energy management activities are accurately quantified and that risks are appropriately allocated.

Economic

Cost savings are very often the primary goal of energy management projects. But there are numerous other ways in which the economic domain influences M&V. The most important is the cost of M&V activities. All stakeholders need to agree on the value of M&V activities.

In this domain, you may need to consult with utility rate/tariff specialists, financial analysts, project finance managers, and accountants – and you will often be expected to provide data and insights that inform investment decisions. You might also work with project managers and financial analysts to identify and quantify risks associated with the project and develop risk mitigation strategies.

You should be aware of the various financing structures available for energy management projects – such as loans, leases, and shared and guaranteed savings contracts.

Another major goal of M&V might be to capture financial incentives and other forms of support, such as tax credits, grants, or rebates. Whatever the case, you will need to exercise professional judgment to obtain the best results within the project's financial constraints.

Energy Budgets

Many organizations have separate budgets for capital and operational expenses, and these budgets often cannot be interchanged. The funds in these budgets come from different sources and have different rules and regulations attached. This can present challenges in M&V activities, especially in energy efficiency projects, where capital investments (e.g., in new and more efficient equipment) can lead to reductions in operational costs (e.g., lower energy bills). You not only need to be able to accurately measure and verify results but also be aware of how the flow of savings integrates with the budgeting process. Your best friend may be the procurement officer.

Utility program incentives may also be framed around either capital or operational costs – or both. Capital cost incentives often take the form of upfront payments, rebates, or tax credits, mitigating the initial investment required for energy efficiency projects or renewable energy systems, such as the installation of solar panels or energy-efficient appliances. The objective is to lower financial hurdles for individuals or businesses contemplating these investments. However, operational cost incentives aim to decrease ongoing costs associated with energy usage. These incentives can manifest as lower electricity rates during off-peak hours, encouraging customers to adjust their usage patterns and reduce peak hour demand. Another strategy involves performance-based incentives, where a utility company may offer a rebate or a reduction on future electricity bills for every kilowatt-hour of energy a customer saves through energy-efficient equipment. These savings are often reflected as credits on the customer's account, reducing the total cost of their utility bill.

Physical

The physical domain of energy management projects refers to the technical aspects of implementing energy efficiency measures that improve performance and accrue impacts. This includes the design, installation, and operation of equipment and systems intended to reduce energy consumption and improve energy efficiency.

In the physical domain, energy management projects typically involve the identification and implementation of specific energy efficiency measures, such as upgrading lighting systems, installing insulation, upgrading HVAC systems, or installing renewable energy systems. These measures may be implemented at various scales, from individual buildings to large industrial facilities or entire communities.

Table 2.4 Energy management activities.

Energy management activities	
Activity	**Responsible role**
Data collection	Auditors (CEA)
Equipment design (sizing)	Engineer (CEM)
Impact targeting prediction	Engineer (CEM)
Installation	Engineer/contractor
Commissioning	Commissioning agent
Functional performance verification	Facilities engineer
Maintenance and replacements	Facilities engineer
Incident reporting	Facilities engineer/safety
Reporting (all levels – utility bills, whole facility, and detailed EMCS logs and trends)	CFO, CEO, and facilities engineer
Removal/salvage	Facilities engineer

Of course, as an M&V professional, you are also responsible for understanding the various technical aspects of M&V projects – that is, the practical application of engineering principles – even if you do not perform all the technical work yourself.

Work in the physical domain starts with the development of an M&V plan that outlines the methods for measurement, the parameters to be tracked, and the project's timeline. Once the plan is in place, you turn to the task of collecting data from a variety of sources, which can include energy meters, building management systems, or even directly from equipment such as HVAC or lighting systems. The information gathered during this stage can relate to energy consumption, system performance, and environmental conditions, among other relevant factors.

Following data collection, you or someone on your team analyzes the data to determine the performance of energy systems and validate the impact of energy management activities. The analyst uses a variety of techniques to accomplish this, ranging from statistical methods to energy modeling. It is then up to you to communicate the results of these analyses to stakeholders in a manner that is understandable and useful.

Although the M&V plays a key role in directing and coordinating activities, the breadth and depth of knowledge required for M&V projects often extend beyond the expertise of a single professional. For this reason, the work of an M&V professional is not done in isolation but is *collaborative*, requiring a diverse team of professionals (Table 2.4).

This team usually encompasses individuals with a wide range of technical skills to ensure that energy efficiency measures are properly designed, installed, and operated to deliver the expected impacts. Types of technical experts you should expect to work with as an M&V professional include:

- **Energy engineers:** They assess the technical feasibility of energy efficiency measures, design energy management projects, and oversee the implementation of energy efficiency upgrades.
- **Contractors:** Contractors install and commission energy efficiency measures.
- **Facility managers:** They are responsible for the operation and maintenance of the energy efficiency measures. In this way, they play a key role in ensuring that the measures continue to function properly over time and deliver the expected energy management impacts.
- **Data analysts:** Data analysts collect and analyze energy consumption data and generate reports.
- **Equipment vendors:** They supply energy efficiency equipment and provide technical support during installation and commissioning.

In addition, many of the elements of M&V overlap with other core energy management activities. For instance, the energy audit contains most of the technical information needed to define a baseline, and commissioning procedures align with functional verification in M&V.

To put it differently, a typical energy management project is a set of processes in the course of which M&V occurs. If we continue to define M&V as a process that leads to a successful settlement, then there will be many points of contact with other players along the way.[17]

2.3 Activities

As an M&V professional, you must call on a range of technical, analytical, and data management skills to collect, analyze, and interpret data. The types of data used in M&V can vary depending on the project or program at hand, but they typically include energy consumption data, equipment performance data, weather data, occupancy data, building envelope data, and maintenance data. Even if you are not *personally* capable of performing all the requisite activities, you should be aware of and have some familiarity with all forms of work associated with M&V.

The primary distinction for M&V activities is between tasks that are performed on site, or "in the field," and tasks that can be performed remotely, such as analytics and general administrative work.

[17]The Facility Manager's Handbook – Joseph Gustin River Publishers.

In the early days of M&V, it was assumed that all projects would require site visits. With the advent of the advanced metering infrastructure (AMI), new approaches to M&V (M&V 2.0) are possible that do not require field work. Much of the data required can be collected and analyzed remotely. However, there are still many situations where field resources are required to get the most out of M&V 2.0 techniques.

Common activities in M&V:

- Baseline development/energy modeling
- Measurement and verification plan development
- Measurement and verification report development
- Calculations of the impacts of energy management activities
- Verification of the impacts of energy management activities
- Calibration of models and systems
- Review and analysis of design documents and specifications
- Trend analysis and statistical analysis
- Training and education
- Quality control and quality assurance
- Risk management
- Continuous commissioning and re-commissioning

2.3.1 Field work

Anyone who has done any work in the field, in real buildings, industrial plants, or manufacturing facilities, knows one thing for sure – they are all unique. No two are piped, wired, or fitted out in the exact same way. This reality both complicates the task of the M&V professional and makes it more interesting (and important). Doing site work to investigate the nature of each project is part of the fun and challenge. It is absolutely required to create the optimal M&V plan for each situation.

There have been attempts to standardize M&V procedures by category of ECM. The FEMP M&V Guidelines and others established procedures for lighting, motors, chillers, etc. But no standard procedure can apply to every situation.

Consider a lighting retrofit in two similar office buildings. Both have the same floor plan and total number of fixtures to be replaced. In one of the buildings, the fixtures are served by a dedicated high voltage (277 V) system. The total lighting load can be measured at the main service panel serving all (and only) the lighting. In such a case, the M&V plan may call for just a few current transformers to capture the entire lighting load. In the other building, the light fixtures are powered from electrical panels that also serve other loads (plugs, etc.). The task of identifying and measuring the lighting circuits in such a building would be orders of magnitude more difficult – and expensive.

Such situations are very common and require an astute awareness of both the retrofit concept and the underlying configuration of the facility's energy distribution systems.

When teaching, I am fond of reminding my classes that in some ways the M&V profession is much more challenging than, say, the medical profession. At least in medicine all the "projects" are wired, plumbed, and ventilated in the same way.

Examples of field work that generally (or always) must be performed on-site include:

- **Site surveys:** This involves visiting the site to identify energy-saving opportunities and inspect equipment.
- **Meter installation:** You – or, more likely, a skilled technician – must physically install meters and sensors to measure energy consumption and performance.
- **Verification and commissioning:** Verification and commissioning usually take place on-site.
- **Calibration and maintenance of meters and sensors:** Meters and sensors can be calibrated in situ.
- **Performance monitoring:** Performance verification activities like conducting a blower door test to measure building envelope air leakage.
- **Measurement tools and approaches:** M&V professionals should be familiar with a variety of measurement tools and approaches, including manual data collection, automated metering systems, data loggers, and other technologies (see Chapter 4).
- **Data management and storage:** M&V professionals should be competent in data management and storage, from database management, to data backup and recovery, to data security. The types of data used in M&V vary from one project or program to another; however, common types include:
    - **Energy consumption data** – on the amount of energy consumed by a facility, as well as data on energy consumption patterns over time.

- **Equipment performance data** – on the performance of specific systems or equipment.
- **Weather data** – on temperature, humidity, and other weather conditions that can affect energy consumption.
- **Occupancy data** – on the number of occupants in a building or facility, as well as data on occupancy patterns over time.
- **Building envelope data** – on the facility's construction, insulation, and other relevant factors.
- **Maintenance data** – on the maintenance and repair history of building systems and equipment.

2.3.2 Analytics

The magic of M&V is the creation of justifiable estimates of the absence of energy use. To accomplish this, the M&V professional will need to be adept at creating, executing, and reporting M&V plans that specify the type of data to be used (field data or utility bills) and, perhaps more importantly, how it will be analyzed. Establishing understanding and agreement among all stakeholders on the exact analysis techniques is the key to turning the magical into the practical.

In Chapter 4, we will provide more details on specific analysis (modeling) techniques. None of them are unique to M&V. They are all common across a range of scientific and engineering disciplines and are covered in any good energy management handbook.

A short list might be:

- **Data analysis:** M&V professionals should have strong analytical skills, including the ability to identify patterns and trends in data, as well as the ability to use statistical analysis software and modeling tools.
- **Utility bill tracking and analysis:** This entails tracking utility bills and analyzing them to identify energy use patterns and trends. It is possible to complete this task off-site by accessing billing data online or through automated software.
- **Energy data analysis:** You can often access data off-site through a remote monitoring system or similar software, to analyze energy consumption data collected by sensors, building automation systems, and other sources.
- **LBNL advanced M&V analysis.**[18]

[18]https://buildings.lbl.gov/emis/assessment-automated-mv-methods

2.3.3 Stipulation

The word "stipulation" can be seen in several contexts in M&V. In its broadest, most extreme sense, stipulation is the use of engineering estimates, with no field measurement.

Stipulating a parameter in a baseline model refers to setting a specific value for a parameter. This may be necessary if there is uncertainty or variability in the data used to develop the baseline model, or if there are specific assumptions or constraints that must be taken into account. For example, if the baseline model is based on historical energy consumption data, it might be wise to stipulate a parameter such as occupancy rates to account for potential variations.

This can help ensure that the baseline model accurately reflects the expected energy consumption or performance of the building. However, the stipulated parameter must be realistic and based on sound assumptions to avoid over- or under-estimating energy consumption or performance.

Stipulating parameters in the baseline model can also act as a risk-sharing mechanism in energy performance contracting. By stipulating certain parameters, the risks associated with those parameters can be shared between the energy service company (ESCO) and the client.

For example, if the baseline model for a building is based on historical energy consumption data, there may be uncertainty or variability in the data due to factors such as changes in operating schedules or equipment. If the ESCO agrees to stipulate a specific parameter, then the risk associated with any changes to that parameter can be shared between the ESCO and the client.

Utility programs often use technical resource manuals (TRMs) to provide standardized and consistent guidelines for evaluating and implementing energy efficiency and demand-side management measures. TRMs are often created using measured values but present "savings" estimates as stipulations.

Technical resource manuals (TRMs):

- **Purpose:** TRMs are comprehensive guides or documents developed by utilities, government agencies, or organizations to provide standardized methodologies and guidelines for estimating the impact of energy management activities, cost-effectiveness, and other relevant factors associated with energy efficiency measures.
- **Content:** TRMs typically include detailed information about various energy-saving measures, such as technologies, equipment specifications,

installation procedures, baseline assumptions, energy impact calculations, and cost data. TRMs may cover a wide range of measures, from simple lighting upgrades to complex HVAC system retrofits.
- **Use:** TRMs are primarily used as reference tools by utility program administrators, energy efficiency implementers, contractors, and customers. They ensure consistency, transparency, and accuracy in evaluating potential energy impacts and cost-effectiveness across different projects and evaluations.
- **Benefits:** TRMs help streamline the evaluation and implementation of energy efficiency projects by offering a standardized approach. They enhance transparency, enable reliable estimation of the impact of energy management activities, support program design, and facilitate reporting and verification.[19]

2.3.4 M&V as counterfactual design

The proper study of mankind is the science of design.

– Herbert Simon

We often think of design as a product, a noun – what has been designed. But according to Herbert Simon, a Nobel laureate in economics, design is better understood as a verb – as the *process* of transforming existing circumstances into more desirable ones. The associated mindset, termed design thinking, encourages unique, stakeholder-focused solutions by employing a blend of logical approaches: deduction (conclusions based on general principles), induction (general theories based on specific facts), and most importantly abduction (best guess explanation for observed facts). It blends qualities such as empathy, personal touch, interpretation, experimentation, synthesis, and optimism. It balances the contrasting demands of creativity and rationality while prioritizing a human-centered approach.

In this sense, M&V is an exercise in design. It is the process of developing and implementing a cost-effective M&V process for any situation. Specifically, the practice of employing design thinking *to develop a counterfactual model* – joining technical expertise with professional judgment and creativity, operating within contextual constraints – can be termed counterfactual design.

To truly be an M&V professional means to be a practitioner of M&V who embraces the concept of counterfactual design.

---

[19]https://www.energy.gov/sites/default/files/2021-09/emv-trm-report-summary.pdf

2.3.5 Valuation

One of the principal objectives of M&V is to allow stakeholders to estimate the financial value derived from energy management investments. While a substantial portion of M&V focuses on the tools, methods, and procedures required to compute the energy impacts, assigning a financial value to these impacts presents a new layer of complexity.

The financial savings resulting from energy management investments are calculated based on the counterfactual approach: what would the bill have been in the absence of the interventions? Answering this question satisfactorily requires the agreement of all stakeholders on the valuation of a hypothetical baseline, and this becomes complicated when energy rates are based on the amount consumed, as the marginal price may fluctuate. Therefore, a thorough understanding of rate structures, avoided costs, and time-of-use implications is imperative.

- *Rate structures* refer to the specific pricing arrangements and tariff designs that determine how energy usage is billed or charged by utilities or energy providers. Rate structures can vary widely and are often tailored to different customer segments or types of energy consumption.[20]
- *Avoided cost* is the monetary value of the energy "saved" by implementing an energy efficiency measure. It represents the cost that would have been incurred if energy consumption had not been reduced. Avoided cost is calculated by multiplying the energy "savings" achieved by the project by the cost per unit of energy. For example, if an energy efficiency measure reduces electricity consumption by 1000 kilowatt-hours (kWh) and the cost of electricity is $0.12 per kWh, the avoided cost would be $120 (1000 kWh × $0.12/kWh).
- The concept of *time of use* refers to the variation in energy costs at different times of the day, influenced by supply and demand. It is an important consideration in the valuation of energy management investments, as energy "savings" achieved during peak demand periods could have higher financial value than those achieved during off-peak periods.

Carbon

As an M&V professional, you play a critical role in ensuring that all stakeholders share a common understanding of the conversion process from the

[20]https://www.riverpublishers.com/aeebook/book_details.php?book_id=901 River Publishers Library

underlying energy impacts to financial units. One aspect to consider while carrying out this conversion is the value associated with carbon reduction.

Carbon reduction is often viewed as a co-benefit of energy management activities. However, including its value in the evaluation of energy management investments can significantly increase the overall worth of these investments. Beyond the immediate financial gains, carbon reduction efforts can also lead to non-financial benefits, such as improved environmental stewardship, regulatory compliance, and an enhanced image vis-à-vis social responsibility.

The challenge, however, lies in assigning a financial value to carbon reduction. There are several mechanisms currently in place for the valuation of carbon reduction through energy management, including:

- **Carbon offset programs:** These initiatives allow companies to "offset" their carbon emissions by investing in projects that reduce or remove carbon from the atmosphere, such as renewable energy or forestry projects.
- **Carbon pricing:** Governments and regulatory bodies have introduced carbon pricing mechanisms to incentivize companies to reduce their carbon emissions. Companies can attain substantial financial savings by reducing carbon emissions and, thus, avoiding or reducing carbon pricing costs.
- **Green financing:** Financial institutions are increasingly offering green financing options for sustainable projects and initiatives, such as green bonds or green loans. These options often provide better terms and lower interest rates compared to traditional financing, making energy management investments more attractive.
- **Corporate social responsibility (CSR):** By incorporating sustainability and CSR into their business strategies, companies can show their commitment to reducing their carbon footprint. This not only enhances their reputation among potential customers but also attracts socially responsible investors.

Recognition of the importance of carbon reduction is steadily rising among organizations investing in energy management. Far from being a peripheral concern, it is fast evolving into a core element in financial assessments of energy management investments.

2.3.6 Non-energy benefits

Energy management activities have benefits that extend beyond reductions in energy consumption or carbon footprint. These non-energy benefits of energy management practices provide indirect values. Some typical non-energy benefits include:

- **Improved indoor air quality:** By implementing energy-efficient measures like improved ventilation, air filtration, and humidity control, indoor air quality can be enhanced, leading to improved health, comfort, and productivity.
- **Increased equipment lifespan:** Energy management practices like equipment maintenance, cleaning, and upgrades can improve the performance of HVAC and other equipment, extending their lifespan and reducing replacement costs.
- **Enhanced occupant comfort:** Energy-efficient measures like improved temperature control or lighting quality can create a more comfortable and productive work environment, leading to improved employee satisfaction and reduced absenteeism.
- **Reduced maintenance costs:** Energy-efficient measures like lighting upgrades, improved HVAC maintenance, and building automation systems can help reduce maintenance costs by allowing for the identification and resolution of issues before they become major problems.
- **Improved asset value:** By implementing energy-efficient measures, buildings can become more attractive to tenants and buyers, increasing their value and reducing vacancy rates.

Several resources offer guidance on how to quantify non-energy benefits, such as the Lawrence Berkeley National Laboratory's webinar materials on *Evaluating and Quantifying the Non-energy Impacts of Energy Efficiency*.[21]

Operations and Maintenance as "Savings"

One category of savings is operations and maintenance (O&M). Effective O&M can result in significant energy impacts. In fact, O&M programs implemented to improve energy efficiency can reduce a facility's energy costs by up to 20% – no capital investment required.

The U.S. Department of Energy's *Operations and Maintenance Best Practices: A Guide to Achieving Operational Efficiency* emphasizes the importance of effective O&M to energy management. It provides methods for calculating the energy efficiency impacts of O&M practices. You can find the latest release of the guide on the Department of Energy's website.

[21] https://emp.lbl.gov/publications/evaluating-and-quantifying-non-energy

2.3.7 The M&V plan

In science, a hypothesis is a prediction that can be tested through experimentation or observation. It is a tentative explanation of some phenomenon, subject to testing, evaluation, and revision.

In energy management, the equivalent of the hypothesis is the estimated impact (or "savings") of an intervention, as stated in an energy audit. Although an energy audit consists of practical *recommendations*, both a scientific hypothesis and an audit's "savings" projections demand a systematic and scientific approach to verification.

The main objective of M&V is to quantify the impacts of energy management activities, not prove or disprove an energy audit's projections. Still, the implied "target" for the performance of the system after implementation (during the reporting period) is also of interest – and can be tracked.

In science, the researcher designs an experiment. In M&V, the M&V professional drafts an M&V plan. This plan must be complete, but it must also reflect contextual constraints – most notably, restrictions on cost. But there are also many other practical considerations, as we will discuss in Chapter 4.[22]

2.4 Definition of a Professional

Your job as an M&V professional is to understand and apply general concepts in a way appropriate to the context and responsive to stakeholders' needs. Once you have analyzed the context, you must develop an M&V plan. In some cases, you will need to create a totally bespoke, custom plan. In other cases (in programmatic contexts, especially), you may be expected to adhere to strictly defined procedures. But you may also face vague, ambiguous, or incomplete guidance. Good luck! Then comes implementation: from assembling and managing a team to arriving at a successful settlement.

Recognizing and responding appropriately to nuances in context from one job to another is the first thing that separates an M&V professional from a technician. Some examples include:

- M&V that is appropriate for the objectives and risk tolerance of the parties to a contract
- M&V that is appropriate to the value of the project and its impact
- M&V that matches the time frame of the guarantees

[22]In many utility programs, there is no site-specific plan for each project. We will discuss later whether this can be considered "real M&V."

- M&V that is appropriate to the available resources and capabilities of those who will conduct the M&V

A professional can navigate the complexities of the situation and, applying judgment, arrive at a satisfactory solution. The technician is called upon to perform the tasks and functions required to execute the M&V plan but is not typically asked to exercise judgment.

2.4.1 Distinctions between professionals and technicians

Like an architect or project manager, an M&V professional must understand a project's status and goals and facilitate communication among all relevant stakeholders. As noted earlier, energy management projects touch on many domains – and you should be conversant (if not an expert) in each of them (Table 2.5).

2.4.2 The role of judgment

Aside from sensitivity to context, a primary difference between the M&V professional and the technician is that the M&V professional must rely on judgment in decision-making.

Professionals in many fields are called upon to use judgment, whether it be diagnosing a medical condition, evaluating the feasibility of a business project, or settling complex legal cases. In all these examples, professionals draw on their knowledge, experience, and critical thinking skills to arrive at informed decisions that reflect the best interest of their clients or the organizations that they represent.

Sometimes, professionals come to different conclusions despite working with the same information and using similar methods. This can happen for a variety of reasons, such as differing interpretations of data, varying levels of expertise or experience, or different decision-making frameworks or approaches. In these cases, it can be difficult to determine who is correct. The M&V professional must recognize this basic uncertainty in advance and establish a communication environment in which it is possible to discuss and justify interpretations and come to mutual agreement among stakeholders. If feasible, a different M&V professional should represent each stakeholder. A healthy respect for client advocacy on all sides will enhance the overall M&V process.

Remember our mantra: savings are defined, not measured. Different methods and perspectives generate different answers – not least of all when it comes to quantifying the impacts of energy management activities

Table 2.5 The roles of professionals and technicians in M&V.

Area	**Technician/analyst's role**	**Professional's role**
Data analysis	Conduct analysis using defined tools, such as Excel, Python, or R	Know which tools to apply
Engineering	Apply energy engineering techniques	Know the limits of engineering analysis
Energy modeling	Creating/calibrating models	Using developed models for various analytical purposes
Value/tariffs	Calculate value using utility rate structures	Handle uncertainty in rates and tariffs
Field data	Perform field metering	Determine whether, when, and where to place meters
Statistical analysis	Perform regression analyses, using languages such as R or Python	Select the appropriate model form: 2/3/4/5 parameters, Bayesian, or frequentist
Regulatory issues	Comply with regulations as communicated by M&V professional	Understand and communicate practical implications of regulations.
Contractual/programmatic requirements	Follow the terms of the M&V section	Develop terms and ensure compatibility with other provisions
M&V plan	Execute specified tasks	Create the task list
Cost of M&V	Provide metering options	Develop budget appropriate to the project requirements
Ethics	Perform functions properly	Adhere to professional ethical standards
Customer relations	Follow rules of access to facilities	Act in the best interest of the client, while following all ethical standards
Safety	Follow all rules	Know all rules

and assigning them monetary value. There is no way to reach a wholly "objective," definitive conclusion.

As an M&V professional, you might find some relief in the fact that every professional endeavor requiring judgment encounters the same issue. Daniel Kahneman captures this general dilemma well in his book *Noise*:

"Wherever there is human judgment, there is noise." Since "noise" of this kind is inevitable, the best way to manage it is to communicate it at the outset, develop a credible M&V plan (agreed upon by all stakeholders), implement the plan as designed, document any necessary adjustments, and report the results clearly.

2.4.3 The guiding principles of M&V

Given the role of judgment in M&V and the impossibility of any single "objectively correct" solution to a problem, you should always consider these "guiding principles" of M&V in your decision-making:

- defer to relevant communities of practice;
- be transparent;
- foster collaboration;
- communicate professionally and effectively.

The guiding principles of M&V can vary depending on the specific context and project being measured, but here are some general principles:

- **Accuracy:** M&V should strive for the highest possible level of accuracy in the measurement and verification of energy impacts. This involves the use of appropriate measurement techniques and tools, as well as rigorous data analysis and verification.
- **Transparency:** The M&V process should be transparent and clearly documented so that stakeholders can understand and verify the results. This includes providing clear explanations of the methods used, assumptions made, and data sources used in the analysis.
- **Consistency:** M&V should be consistent over time so that energy impacts can be reliably measured and compared across different projects or time periods. This requires the use of consistent measurement methods and data collection procedures.
- **Objectivity:** M&V should be conducted in an objective and unbiased manner to ensure that the results are reliable and credible. This may involve using third-party M&V providers or establishing independent review processes.
- **Relevance:** M&V should be relevant to the specific project being measured, taking into account the project goals, context, and constraints. This requires careful consideration of the appropriate measurement and verification methods to use.

- **Cost-effectiveness:** M&V should be conducted in a cost-effective manner so that the benefits of the energy efficiency project are not outweighed by the cost of measuring and verifying the impact. This requires balancing the need for accurate measurement with the resources available for conducting the M&V.

IPMVP describes the fundamental principles of M&V as follows:

The fundamental principles of good M&V practice are described below, in alphabetical order.

- **Accurate:** M&V reports should be as accurate as the M&V budget will allow. M&V costs should normally be small relative to the monetary value of the impact being evaluated. M&V expenditures should also be consistent with the financial implications of over- or under-reporting of a project's performance. Accuracy tradeoffs should be accompanied by increased conservativeness in any estimates and judgments.
- **Complete:** The reporting of the impacts of energy management activities should consider all effects of a project. M&V activities should use measurements to quantify the significant effects, while estimating all others.
- **Conservative:** Where judgments are made about uncertain quantities, M&V procedures should be designed to under-estimate the impacts of energy management activities.
- **Consistent:** The reporting of a project's energy effectiveness should be consistent between:
    - different types of energy efficiency projects;
    - different energy management professionals for any one project;
    - different periods of time for the same project;
    - energy efficiency projects and new energy supply projects.

"Consistent" does not mean "identical," since it is recognized that any empirically derived report involves judgments, which may not be made identically by all reporters. By identifying key areas of judgment, IPMVP helps to avoid inconsistencies arising from the lack of consideration of important dimensions.

- **Relevant**: The determination of impact should measure the performance parameters of concern, or least well known, while other less critical or predictable parameters may be estimated.
- **Transparent**: All M&V activities should be clearly and fully disclosed.

2.4.4 Community

Local Communities

M&V's fundamental concepts know no borders. Their underlying mathematics and physical principles are universal. However, the practice of M&V can vary greatly based on local needs, proclivities, and culture.

The context matters. The community of practice matters.

When disputes arise that cannot be easily resolved, defer to the relevant professional community standards. These standards define what is considered acceptable and appropriate within the community. Professional community standards can also dispel mistrust by providing a framework for assessing and resolving disputes. For example, in M&V, we must base our conclusions on sound evidence and reasoning and disclose any potential conflicts of interest.

One prominent example of successful community standards comes from South Africa, where Karel Steyn documented decades of experience.[23] Some of these lessons were integrated into Energy Performance Measurement and Verification: Guidance on Data Quality.[24]

M&V's Global Reach

But because energy management is practiced nearly anywhere energy is utilized, the M&V professional must also be aware of the international standards and protocols of our profession. Thanks to this guidance, a competent M&V professional, even when introduced to a new context, should be able to determine what the specific needs of the context are and how to apply the fundamental M&V process.

In my experience, many localities either have or want to have their "own" protocol. However, among the dozens of existing "local" protocols, very few contain unique or conflicting methods.

It is for this reason that the Association of Energy Engineers has been able to leverage a group of experts from around the world to conduct a job task analysis and develop an internationally relevant "body of knowledge" for its Certified M&V Professional (M&V professional) accreditation.

---

[23] "Lessons Learned – Energy Measurement and Verification – The South African Version" April 2021.

[24] https://www.oneplanetnetwork.org/sites/default/files/energy_performance_measurement_and_verification.pdf

As an M&V professional, you should be able to assemble a team anywhere in the world that can perform M&V successfully and in accordance with best practices.

2.4.5 Transparency

In many cases, multiple parties are involved in M&V, including facility owners, energy service companies, utilities, and third-party evaluators. Each party may have different goals, objectives, and perspectives, which can lead to conflicts and disputes if not addressed early on.

For this reason, it is critical to demonstrate transparency by providing all parties with access to the same data on energy consumption, equipment performance, and other key metrics. Likewise, open channels of communication can keep all parties on the same page and help identify and resolve issues before they become major points of contention.

Detailed documentation, well-developed M&V plans, complete explanation of underlying assumptions, and referenced standards/guidance will all demonstrate a commitment to transparency.

2.4.6 Collaboration

Collaboration is always important in M&V, but especially in complex M&V projects involving multiple parties. Everyone should be working toward the same goals – sharing best practices, identifying areas for improvement, and coming together to find solutions to difficult problems. Even when M&V professionals take the responsibility of representing their clients' interests – and those interests are opposing – all parties succeed when the M&V process proceeds smoothly.

2.4.7 Communication

Of all the guiding principles of M&V, communication is perhaps the most important for maintaining the credibility of the M&V process, avoiding disputes among stakeholders, and arriving at a successful settlement. Always keep in mind the following:

- Consider your audience, especially when communicating with multiple stakeholders at once. M&V often involves highly specific terminology that will not be understood by all parties. The contractual terms are important to the lawyers but perhaps not to the facility engineers – and vice versa. The statistical review of uncertainty will likely go beyond most participants' understanding. Despite these difficulties, as an M&V professional, you must communicate in such a way as to foster trust and cooperation. As Nobel laureate Herbert Simon argued, communication is not just about transmitting information – it is also about building common understanding and creating shared mental models. This is particularly important in M&V, where multiple stakeholders from different domains and professions are involved in the project.

- You should employ a variety of communication methods, from meetings, to emails, to formal reports, to keep stakeholders informed. And do not limit yourself to spoken and written language – you can use visual aids, such as graphs and charts, to communicate complex data in a more accessible way.
- While much of your communication will be informal, at some point, you will generate a formal M&V plan that captures the agreement among stakeholders as to how the M&V will be performed and reported. This plan may be embedded within a larger document that lays out all the terms and definitions applicable to the political, financial, technical/physical, and contractual domains, and it is likely that the plan will refer to a specific M&V guidance document.
- Be proactive as a communicator – do not wait for stakeholders to come to you with questions or concerns.

2.4.8 Dispute resolution

When M&V projects lack good communication, disputes soon follow. Of course, the best method of resolving disputes is to avoid them in the first place, and an M&V professional should be an expert in dispute avoidance. Still, some disputes may arise despite your best efforts.

By following the guidelines below, you can improve your chances of avoiding disputes:

- **Clearly define the objectives of the M&V plan and communicate them to all stakeholders:** This includes defining the scope of work, the performance metrics, and the roles and responsibilities of each stakeholder.
- **Use a common language to communicate technical information:** M&V professionals should avoid using technical jargon and use simple language that can be easily understood by non-technical stakeholders.

- **Communicate regularly with all stakeholders:** This can include regular meetings, progress reports, and updates on the M&V plan.
- **Address concerns or issues raised by stakeholders in a timely and transparent manner:** This helps to build trust and credibility with stakeholders and ensures that any issues are addressed before they become larger problems.
- **Provide clear and concise reports that accurately reflect the results of the M&V plan:** These reports should be easy to understand and provide sufficient details to allow stakeholders to make informed decisions.
- **Be open to feedback from stakeholders and be willing to make changes to the M&V plan as needed:** This can help to ensure that the M&V plan is aligned with the needs and expectations of all stakeholders.

3

Standards of Practice

In this chapter, we will look at the major documents that (attempt to) define "M&V" at a fundamental level.

In a way, this should be a simple chapter: each set of printed words and equations – the results of experts meeting over many years – constitute an agreement at a particular time about the meaning of "M&V." However, these documents have all undergone revisions, and as an M&V professional, you should be aware of the differences between the documents and the versions of each.

Remember, these documents were mostly the product of committees. They were mostly committees of volunteers. They necessarily contain large and small inconsistencies that neither time nor inclination was sufficient to resolve. It is the job of the M&V professional to "read between the lines," see the bigger picture, and learn how to interpret each document.

Ultimately, these standard documents are imperfect road maps. As an M&V professional, you should grasp the primary purpose of each, their structure, and their usefulness within various contexts. But remember, the fundamental theory of M&V can be applied *without reference to any specific document.*

3.1 Terminology, Taxonomy, Definition, and Specification

Before delving into the history and details of the actual documents, let us address the issue of terminology, taxonomy, definition, and specification (see definitions at the end of this chapter). Perhaps nowhere are these concepts more overlapping and confusing than in these standard guidance documents.

We can define the word "terminology" as the body of specialized or technical words, phrases, and expressions used in a specific field, profession, or community – continually revised by practitioners and community members to meet the emerging needs, complexities, and advancements of the field. Here, the important takeaway is that organizations continually redefine and adapt terminology to suit their purposes. The goal of the writers of each of the M&V (and EM&V) protocols was to establish terminology sufficient to guide users on how to quantify the impacts of energy management activities, in both project and programmatic contexts. The resulting terminology offers useful approximations, not precise definitions.

The concept of "options" in many of these documents gives perhaps the clearest example of this. When we wrote the original FEMP and IPMVP protocols, we needed general terminology to distinguish between broad categories of approaches to performing M&V. We coined the term "options" to provide an easy way to communicate some fundamental distinctions. But the resulting categories were necessarily broad and were insufficient as definitions to capture all the nuances involved in developing a site-specific M&V plan. Over time, users adopted their own interpretations of the general "options" framework. And the M&V documents have been edited to attempt to lay precise definitions onto general categories (the distinction between Option A and Option B in the FEMP guidelines and IPMVP has shifted over time) with little success. Later in this section, we will discuss the history of the development of the "options" terminology.

Another bit of terminology can evoke contentious discussion. Can "savings" be "measured"? The title of ASHRAE Guideline 14 is "Measurement of Energy, Demand and Water Savings." The foreword to the 2023 version of this document explains:

Determining savings*with measurements in accordance with this guideline involves measuring post-retrofit energy use and comparing that to the measured pre-retrofit use, adjusted or normalized, to act as a proxy for the conditions that would have prevailed had the retrofit not been performed. Therefore,* ***determining energy savings*** *through the use of measurements involves more than just verifying that new equipment has been installed and can function as expected, although those tasks are usually a necessary part of determining savings. In addition,* ***energy savings cannot be claimed to be measured*** *if no pre-retrofit data are available.*

I agree with the use of the word "determine." I do not think one can ever claim to measure the absence of something.

Another fundamental issue is the inconsistent terminology distinguishing "suggested activities" (i.e., *shoulds*) from "compliance with standards" (i.e., *musts*). A range of terms are used to imply required activities: Best practice?

Suggested practice? Adherence, conformance, compliance? Each of these words can have a different meaning. And each of the main M&V documents takes a slightly different approach. In general, if the M&V is part of a contract, look to the contract terms to define how it fits in. If the M&V is part of a program, read and follow the rules of the program.

One thing is for sure – different interpretations of terminology can lead to disputes. The M&V professional needs to ensure that all stakeholders in their context maintain a common understanding of the role of the guidance documents and agree to common definitions of key terms.

3.2 A Brief History of the Genesis and Status of "M&V"

As it is generally understood and practiced now, M&V is the process of quantifying the impacts of energy management activities using data collected in the field. The nature of the project, the needs of stakeholders, and the available budget determine the type and amount of data that must be collected and analyzed. At one end of the spectrum, M&V can be a complex and expensive undertaking. At the other end, the M&V plan might be "just measure something." Not all projects can afford custom M&V, and many do not require it.

For as long as energy has been utilized, there have been practical attempts to understand and manage its use. The use of field data to analyze the energy use of facilities goes back to at least the 1800s. The development of energy models for this purpose began in the 1960s.

Common-sense engineering drove most energy system designs through the 1960s. But the 1970s saw a dramatic increase in energy awareness. Oil supplies throughout the world were impacted by the Arab oil embargo in 1973–1974, which precipitated widespread energy shortages and high energy prices in many countries. The energy crisis led to a renewed focus on energy efficiency and conservation, as businesses and governments sought to reduce their energy consumption and save on energy costs. This in turn led to the development of energy management programs, which aimed to help businesses and organizations monitor and control their energy usage.

As energy management programs became increasingly common, there was a growing need for reliable methods of measuring and verifying their impacts. In the late 1980s and early 1990s, two energy management concepts arose that required a quantitative assessment of results: utility programs and performance contracting.

3.2.1 Program EM&V

Utility programs, often called demand-side management (DSM) programs, emphasized energy efficiency and the adoption of new technologies, such as variable speed drives, energy efficiency motors, and lighting systems. Major developments in DSM programs during the 1980s include:

- **Building energy codes:** The decade saw an increased focus on building energy codes, which set minimum energy efficiency standards for new buildings and major renovations.
- **Advances in energy efficiency technology:** Advances in lighting and HVAC technology.
- **Time-of-use pricing:** Utilities began to adopt time-of-use pricing, charging higher rates during peak demand periods to encourage consumers to reduce their energy use during these times.

California was an early leader in the development of DSM programs, as well as in the development of tools for evaluating their effectiveness – generally known as evaluation, measurement, and verification (EM&V). New Jersey was another early leader in the development of protocols to support utility-scale energy management programs.

Today's EM&V community is supported by several organizations that facilitate communication on core issues related to quantifying the impacts of energy efficiency programs. Among the most prominent are the International Energy Program Evaluation Conference (IEPEC) and the Association of Energy Service Professionals (AESP). Meanwhile, current EM&V guidance documents include the *California EM&V Framework*, the *California EM&V Protocols*, the *Energy Efficiency Program Impact Evaluation Guide*, and the State and Local Energy Efficiency Action Network's *Energy Efficiency Program Impact Evaluation Guide*.

3.2.2 Energy savings performance contracting

Energy savings performance contracting (ESPC) evolved alongside DSM programs to help organizations finance and implement energy efficiency projects. ESPC contracts became more common in the 1980s as utilities and governments tried to encourage energy efficiency and conservation.

Nowhere was this more evident than in the U.S. federal government, which passed the Energy Policy Act (EPACT) in 1992. This law authorized the use of ESPC in federal facilities. And while it did not provide specific guidance for the quantification of results, it did indicate the need for a general methodology.

Inset: Relevant Definitions from Code of Federal Regulations (CFR) § 436.31

- *Energy conservation measures* means measures that are applied to an existing federally owned building or facility that improves energy efficiency, are lifecycle cost-effective, [...] and involve energy conservation, cogeneration facilities, renewable energy sources, improvements in operation and maintenance efficiencies, or retrofit activities.
- *Energy baseline* means the amount of energy that would be consumed annually without implementation of energy conservation measures based on historical metered data, engineering calculations, submetering of buildings or energy consuming systems, building load simulation models, statistical regression analysis, or some combination of these methods.
- *Energy savings performance contract* means a contract that provides for the performance of services for the design, acquisition, installation, testing, operation, and, where appropriate, maintenance and repair of an identified energy conservation measure or series of measures at one or more locations.
- *Energy unit savings* means the determination, in electrical or thermal units (e.g., kilowatt-hour (kwh), kilowatt (kw), or British thermal units (Btu)), of the reduction in energy use or demand by comparing consumption or demand, after completion of contractor-installed energy conservation measures, to an energy baseline established in the contract.
- *Energy cost savings* means a reduction in the cost of energy and related operation and maintenance expenses, from a base cost established through *a methodology set forth in an energy savings performance contract*, utilized in an existing federally owned building or buildings or other federally owned facilities as a result of:

    1. the lease or purchase of operating equipment, improvements, altered operation and maintenance, or technical services; or
    2. the increased efficient use of existing energy sources by cogeneration or heat recovery, excluding any cogeneration process other than a federally owned building or buildings or other federally owned facilities.

* * *

The need for a standard methodology to be "set forth in a ... performance contract" drove FEMP to seek industry consensus on a set of appropriate methods. The federal ESPC program provided an immediate need for these standards; however, the Department of Energy also saw an opportunity to create general guidance on M&V that would benefit the broader ESPC industry. At the time, there were regional and utility-specific programs and guidance documents in New Jersey (Public Service Electric & Gas's Standard Offer programs) and California (Southern California Edison's ENvest and Pacific Gas & Electric's Power Saving Partners), but no one document satisfactorily fulfilled that need.

So, from 1994 to 1996, the Department of Energy led an effort that resulted in the North American Energy Measurement and Verification Protocol (NEMVP), which later became IPMVP. At the same time, ASHRAE Technical Committee (TC) 7.6 began work on Guideline 14 (the third version of this document was released in 2023).

As the early development of FEMP, NEMVP/IPMVP, and ASHRAE 14 was simultaneous, there was much overlap of membership on the committees. All three documents adopted similar approaches, yet each had a specific focus. And today, all three are still considered the foundational documents of "M&V."

- FEMP's M&V guidelines are specific to the needs of the U.S. federal government contracts and procurement. The major emphasis is compliance with U.S. federal procurement rules.
- IPMVP provides general guidance on some of M&V's basic concepts and best practices. It establishes a general framework, within with users can establish their own applications.
- ASHRAE 14 calls for more scientifically rigorous engineering methods but still offers practical guidance on how to comply with the document's specifications. Although it goes into more in-depth detail, it generally follows the IPMVP framework.

More documents, like the ISO 50001 series, Uniform Methods Project (UMP), and others, have been added to the mix over the past three decades. As an M&V professional, you need to be aware that each of these documents (and many others) has been updated over the years. Contracts signed under a previous version may not be in line with current versions, due to revisions in guidelines or changes in specific methodologies, which could impact the performance of a project.

To maintain credibility in the process, you should specify which version or aspect of the protocol that you plan to adhere to. Are you employing the latest revision, or are you referencing a version that was current at the project's commencement? Are you following general principles, or are you applying specific, detailed guidelines from the protocol?

Simply declaring "I follow a protocol" is not enough. This statement lacks the necessary detail and could lead to misunderstandings or misconceptions. IPMVP, for instance, provides a broad framework for verifying the results of energy management projects. However, it has been updated multiple times since its first publication, reflecting changes in technologies, methodologies, and industry best practices.

In general, it is more important for all parties to understand and follow general M&V principles than to claim adherence to any particular M&V protocol.

3.2.3 The evolution of the "options" terminology

Throughout this book, we emphasize that many of the concepts that make up "M&V" are not unique to quantifying the result of energy management. We hope that people with a wide range of backgrounds and training will see how their skills can be applied to M&V projects. When getting started in M&V, it is obvious that the first place to look for guidance would be the industry's standard documents. The M&V protocols discussed in Chapter 3 are attempts to create standardized language for M&V under various contexts. In a positive sense, the terminology adopted in those documents can be helpful in building a common understanding of the core issues facing any M&V project. What type of model should I use? What measurement boundary? How long should I measure? All of these are good questions. We further emphasize that the role of an M&V professional is to understand the tradeoffs and use judgment to answer these questions when crafting an M&V plan.

However, as discussed in Chapter 3, the standard documents are not always entirely clear about every detail. They sometimes contain language that requires further interpretation. It can be confusing even to long-term experts in the field. One reason for this is that people tend to take what they want from the documents and ignore parts that might conflict. For any particular M&V project, all stakeholders on a project need to have the same interpretation.

For most purely technical topics, interpretation is not typically an issue. There is less disagreement about how to take measurements and readings than there is about whether to take them and for how long.

But with two areas, the M&V professional is advised to be very careful in reading, interpreting, and communicating the standard documents. These are the terminology of "options" and "adherence."

In Chapter 2, we made the case that M&V relies on two types of system impacts – performance and "savings." We emphasize that performance can always be measured, but "savings" can only be estimated by applying the counterfactual method.

This distinction is key to understanding how the options terminology arose. And if you fully understand it, you will be able to navigate a topic that confuses most experts to this day.

Let us take a deeper look at the challenges of creating useful terminology, categories, definitions, and specifications that attempt to guide a professional in M&V – and at the end, you will find a proposed set of definitions for these terms.

* * *

Note

In the language of the FEMP and IPMVP, Option A is for performance, while Options B, C, and D are for "savings."

* * *

A Brief History of "Options"

I was first exposed to M&V when I was asked to take over a pilot project using ESPC at the Lawrence Berkeley National Laboratory in 1992. This pilot was one of three projects that the Department of Energy had authorized to learn how to apply ESPC using federal and local procurement rules. There were no protocols or guidance available; so we had to apply common sense and solid engineering principles.

We began the project with a big budget for metering and monitoring. We had access to data loggers, an energy management system, and any tool we might need for measurement of the systems. Our colleagues at Pacific Northwest National Laboratory had established a very detailed baseline. But as the project unfolded, we began to work more closely with the ESCO implementing the projects. We began to see that the building operation (a laboratory) was not consistent, and hence, the predictions of "savings" were highly dependent on the types of experiments (and hence the operation of the building) planned for the coming years. We noted that lighting loads varied widely depending on the schedules of the academic year and research needs. The concept of risk and responsibility came to the forefront over more technical and less contentious issues. The proposed lighting system's "performance" would not be difficult to verify. But the "operation," the number of hours a year, was anyone's guess. Similarly, the "performance" of the variable speed drive on a pump could be verified. But the number of hours of operation would be hard to predict and would create risk for the ESCO.

As stated in the final report:

> To control risks, and reduce costs, many ESCOs propose baseline methodologies that essentially stipulate the energy savings—eliminating the need for expensive metering and analysis. This can result in a more comprehensive package of measures, including risky measures, as well as measures that are difficult/expensive to baseline. Such an arrangement transfers risk to the owner. In these situations, it is critical for the owner to obtain independent expert advice. Further, if performance is not measured, inspecting and testing procedures should be specified that will ensure the measures are working throughout the life of the contract.

When it came time to finally agree on how we would verify the lighting system "savings," we sat down in a conference room and came up with this solution. (Note that in this building, all the lighting was on one set of higher-voltage distribution panels. We could get a reading of lighting power at any time from measuring the amperage and voltage on just three wires.) We turned on all the lights in the building and took a reading. After the retrofit, we did that again (after assuring lighting quality). The difference constituted the increase in performance for the system. We then agreed upon an average number of hours of operation – aka equivalent full load hours – based on the detailed baseline data we had collected.

We measured the main variable of interest (performance) and "stipulated" the other (operation). I prefer to call this measured performance, stipulated operation. And as we will see, that is what became the initial definition of "Option A." But there is a bit more to the story.

After finishing this pilot program, I was invited to lead an effort for the FEMP program to take what we had learned and leverage it to create a standard methodology for the new FEMP ESPC program, authorized by EPACT in 1992.

My first stop on the FEMP M&V project was the ASHRAE winter meeting in New Orleans in 1994. There I learned that ASHRAE had just initiated work on proposed guidelines for "measuring energy savings": "14P." What could be better? My plan became to work with the Technical Committee 7.6 and have FEMP adopt this new industry standard. Little did I know that ASHRAE's clock speed did not match FEMP's needs. At the June summer meeting in 1994, I learned that it was going to take years to complete 14P. And I had six months to come up with an M&V "protocol."

At this point, the Department of Energy had a broader interest in a national standard for M&V, to cover all commercial ESPC and energy efficiency program needs. Later in 1994, we convened a series of meetings with a broad range of national stakeholders to set in motion a process to develop a national standard in less time than ASHRAE.

The next couple of years were a hectic blend of technical and political jousting about what constituted "proper" M&V. Good ideas came from the experience of many different groups. But the needs of these groups often differed. Some saw M&V as a scientific endeavor, with the goal of establishing "true savings." Others (myself included – from my recent experience) saw M&V as a negotiation with principled actors to arrive at a satisfactory solution on a case-by-case basis. In addition, some groups were dead set on using whole facility methods while others had vested interests in using field measurement of end-use technologies. The question of how much stipulation to allow was another source of passionate discussion.

Months of discussion frequently ended in disagreement when we attempted to arrive at a single methodology that would satisfy everyone. Finally, on a conference call, after we had beaten the topic to death for the *n*th time, I made a "suggestion" that we stop arguing and just agree that our guidance would include "options." We would call them A, B, and C for simplicity. There was silence for what seemed like a couple of minutes, followed by a muted "Okay." Then, "Sure." Then, "Guess so." The "options" terminology was officially born.

At that moment, the three categories that comprised Options A, B, and C were roughly:

- **Option A:** As with the Building 62 pilot project, Option A measured "performance" of an ECM with stipulated "operation." This could be applied to any system that had a measurable performance characteristic.
- **Option B:** This entailed submetering of end-use load measurements, which always requires installing or otherwise utilizing meters on the actual equipment in the retrofit.
- **Option C:** This was the whole facility method – utility bill analysis or analysis of meters installed on the whole facility.

The adoption of these "options" was essentially a high-level categorization. The clearest distinction in possible methods was between "whole facility" – aka utility bill analysis – and end-use metered techniques (requiring field measurement). As it turned out, ASHRAE 14 maintained that distinction but never adopted a pure performance option like Option A.

In the first IPMVP document (1996), this distinction was clear, though using different terminology (measured capacity, stipulated/measured consumption). The 1997 version of IPMVP used the same definitions for Options A, B, and C but added an Option D, calibrated simulation. This brought IPMVP closer in alignment with ASHRAE 14, which also included calibrated simulation.

Figure 3.1 From the 1997 IPMVP Table of Contents.

As can be seen in the table in Appendix I, subsequent versions of IPMVP adopted new definitions. And these definitions evolved over time. They continued to evolve with a new definition for Option B in the 2022 version of *IPMVP Core Concepts*.

Has anything changed in the core concepts of the scientific method? Of energy engineering? Of risk management? I do not think so. What changes with each version is the interpretation as espoused by the most recent group of volunteers.

Let us back up a bit by looking at two final points: categorization and the role of standards.

As we saw in Chapter 2, the concepts that comprise M&V revolve around the process of using collected evidence to assess the impact of energy management. And a critical constraint is that any process deployed must be cost-effective.

We saw that while M&V has adopted many of the tools of science, statistics, and econometrics, it must be implemented in the context of risk sharing between parties. It must adapt to the realities of epistemic uncertainty (non-routine events). We make the strongest case possible that the role of the M&V professional is to understand these constraints and work cooperatively with all stakeholders to establish and implement a plan that reflects these realities.

Standard documents can assist in the conduct of good M&V by laying out clear and reasonable frameworks. But they can confuse the issue if they contain logical inconsistencies. Option terminology (categorization) and adherence (compliance) are two areas where this causes much confusion – and they are interwoven.

Categorization and Compliance

Categorization can be very difficult. To be meaningful, there should be little room for interpretation. The original definitions of Options A, B, and C, though crude, essentially met this definition.

Option A allowed for stipulation. Option B was for isolated systems. Option C was for whole facilities. Within that categorization, subcategories could be (and were) further defined. Suggested subcategories included variable vs. constant loads in both or either the baseline or reporting period, monthly vs. hourly data, statistical vs. physical models, etc.

Having established categories, the next step is to specify what constitutes adherence. There are 156 instances of "should" in the 1996 NEMVP. There are 237 in the 1997 IPMVP. In the 2002, IPMVP there are 120 uses of "should," but a new section was added: 3.5 – Adherence with IPMVP.

3.5 Adherence with IPMVP

The IPMVP is a framework of definitions and methods for assessing energy savings. The IPMVP framework was designed to allow users to develop an M&V plan for a specific project. The IPMVP was written to allow maximum flexibility in creating M&V plans that meet the needs of individual projects, but also adhere to the principles of accuracy, transparency and repeatability. In the case where users are required to demonstrate adherence, or wish to claim adherence with the IPMVP, the following issues should be addressed.

- The two parties should identify the organization/person responsible for M&V activities. This organization/person should be responsible for approving the site-specific M&V plan, and making sure that the M&V plan is followed for the duration of the contract.
- The M&V plan should clearly state which IPMVP Option (or combination of Options) and methods (linear regression, multiple regression, bin method etc.) will be used to determine the energy savings.
- The two parties should agree on a site-specific plan that specifies the metering/monitoring to be conducted. The plan should clearly state how the baseyear energy use and baseyear conditions are to be established including: what measurements are to be taken, how the data are to be used, what variables are to be stipulated and the basis for stipulation. The plan should provide information on the metering equipment, its calibration, the location of measurements, duration of the metering period, accuracy of the measurement process, etc.
- The M&V plan should specify the details of how calculations should be made by stating the variables (run-time hours, electrical consumption in a lighting fixture, kW/ton, etc.) that should be measured and any associated assumptions.
- The two parties should agree on how quality assurance should be maintained and replicability confirmed.
- The M&V plan should list the reports to be prepared, their contents and formats, and a stipulated time frame during which they should be furnished.
- All terminology should be consistent with IPMVP definitions.

If you read through these bullets, you will see that "adherence" was essentially a specification for how parties could establish an acceptable M&V plan. Under this definition, the guidance document is more like a glossary with guiding principles, not a rigorous specification of "how to do M&V."

When conducting M&V now, the question still arises, "Who decides?" The whole issue of science vs. practical confirmation of intent remains.

Remember, sometimes a spot measurement is all you can afford. If you can simply confirm that an ECM is working, that constitutes "good" M&V. In other cases, a detailed baseline and precise measurements in the reporting period are "worth it." Only the stakeholders of the project can determine that. My preferred method is to perform an uncertainty analysis (using Monte

Carlo) that attempts to identify the most cost-effective measurements. But that is still considered advanced and not detailed in the protocols.

Nevertheless, in some contexts, you will have to work within established rules:

- The FEMP M&V Guidelines were written specifically to match the contractual requirements of the Department of Energy FEMP ESPC and Super ESPC program, including detailed procurement rules.
- ASHRAE 14 took the path of establishing minimum requirements for compliance.
- IPMVP lives somewhere in the middle – and continues to evolve.

I hope this brief history of the "options" terminology helps you to understand the challenges inherent in writing M&V guidance documents and an appreciation for the nuances of categorization of standards vs. adherence. Perhaps you even gained some compassion for those who have had to wrestle with attempting "core definitions" through a committee process. In any case, you will almost certainly profit from establishing a clear and common understanding of this topic with all your stakeholders prior to engaging in an M&V project.

* * *

The table in Appendix I contains the core definitions of Options A and B over the last versions of IPMVP.

* * *

Taxonomy

- **Definition:** Taxonomy is the systematic classification and organization of concepts, objects, or terms into hierarchical categories based on their similarities and relationships.
- **Purpose:** Taxonomy aims to provide a structured framework for organizing and representing knowledge within a specific field or domain.
- **Key characteristics:** Taxonomy involves the creation of hierarchical categories, often represented as a tree-like structure, to group-related concepts or objects. It focuses on establishing relationships and hierarchies between different categories.

Categorization

- **Definition:** Categorization involves assigning concepts or objects to specific categories based on their shared characteristics or attributes.

- **Purpose:** Categorization is used to systematically organize and group concepts or objects, allowing for more accessible information retrieval, analysis, and understanding.
- **Key characteristics:** Categorization involves classifying and placing concepts or objects into predefined categories. It focuses on grouping similar items based on commonalities or shared characteristics.

Terminology

- **Definition:** Terminology refers to the specialized vocabulary or terms used within a particular domain, discipline, or industry.
- **Purpose:** Terminology aims to establish a shared language and understanding among individuals working within the same field, facilitating effective communication and knowledge exchange.
- **Key characteristics:** Terminology consists of the specific terms, phrases, and jargon that professionals or experts within a field commonly use. It ensures precise and accurate communication by providing a standardized vocabulary for discussing and referring to concepts, objects, or phenomena within the domain.

Definition

- **Definition:** Definition is the explanation or clarification of the meaning of a specific term or concept.
- **Purpose:** Definitions aim to provide clear and precise explanations that outline the essential characteristics or attributes of the term being defined.
- **Key characteristics:** Definitions help establish a common understanding of a particular concept or term by eliminating ambiguity and preventing misunderstandings. They provide concise explanations that convey the term's meaning, scope, and context.

Specification

- **Definition:** Specification refers to specific criteria, attributes, or details that further refine or define a concept, object, or term.
- **Purpose:** Specifications provide additional information or qualifiers to make a concept, object, or term more precise.
- **Key characteristics:** Specifications add specific details, parameters, or conditions to enhance the understanding or application of a concept, object, or term. They provide additional criteria or requirements that help define the subject's boundaries, context, or characteristics.

3.3 Documents

In the early years of the current era of M&V, three documents appeared that still constitute the core M&V concepts. Let us look at each in a little more detail.

3.3.1 FEMP[1]

FEMP's M&V guidelines consist of standard protocols for assessing energy and water management activities within the U.S. federal government context, specifically for federal agencies looking to quantify the impacts of such projects. The guidelines' primary focus is compliance with U.S. federal energy regulations and requirements. They are aimed at ensuring that federal agencies not only meet their energy management goals but also adhere to federal procurement rules for energy savings performance contracts.

FEMP also offers a risk and responsibility matrix to identify and allocate risks and responsibilities among project stakeholders. The matrix helps define who is responsible for each aspect of the project and what risks each party is willing to assume. This is important for managing the project effectively and ensuring that all parties have a clear understanding of their roles and responsibilities. The matrix can also assist in identifying potential risks and issues before they become major problems.

3.3.2 IPMVP[2]

The purpose of the International Performance Measurement and Verification Protocol (IPMVP) is to provide a framework for planning, implementing, documenting, and reporting the impacts of energy efficiency, water efficiency, and renewable energy projects. The protocol outlines the principles and concepts of M&V and provides guidance on M&V planning, M&V options, and data analysis methods.

Unlike FEMP, IPMVP does not have a specific risk matrix, though the 2022 version of IPMVP does mention risks and responsibilities multiple times. Instead, IPMVP emphasizes the importance of a comprehensive M&V plan detailing the M&V process to be followed for a specific project.

IPMVP acknowledges that there are inherent uncertainties and risks in M&V and that the level of risk will vary depending on factors such as the

[1] https://www.energy.gov/node/1413841
[2] https://evo-world.org/en/products-services-mainmenu-en/protocols/ipmvp

complexity of the project, the quality of the baseline data, and the reliability of the measurement equipment. Therefore, the protocol encourages practitioners to develop M&V plans that address specific project risks and to use appropriate risk management tools and techniques, such as sensitivity analysis and Monte Carlo simulations, to quantify and manage risk – at a reasonable cost.

> IPMVP (and FEMP) contains an M&V "Option" that is based on short-term performance measurements – "Option A."

Option A allows for the measurement of a single or small set of performance parameters, like lighting or motor wattage. It does not necessitate the creation and maintenance of a full counterfactual model, and as such, it is often quicker, easier, and less expensive to implement than other M&V methods.

If you tend to be cynical, you might see the success of IPMVP stemming from the validation of the simplest method, "Option A." This method offers an attractive route for organizations seeking to meet their M&V obligations without incurring substantial costs or needing specialized modeling expertise.

3.3.3 ASHRAE 14[3]

ASHRAE Guideline 14-2014 is a conformance framework for measuring and verifying the impacts of energy management activities (the most recent iteration was released in 2023, while this book was still being written). It aims to standardize and provide a consistent approach to M&V and to ensure that the reported impacts are credible and transparent. The guideline also gives guidance on the selection of appropriate M&V options based on project characteristics and objectives.

From the Foreword

Guideline 14 was developed by ASHRAE to fill a need for a standardized set of energy, demand, and water savings calculation procedures. The intent is to provide guidance on minimum acceptable levels of performance for determining energy and demand savings, using measurements, in commercial transactions. It is entirely possible to have a sale/purchase, lease, or other arrangement for energy-efficient equipment that does not involve measurements. Indeed, the vast majority of transactions are of this type. However, if the savings determination is to be based on measurements, certain minimum

[3]https://watchwire.ai/ashrae-guideline-14/

requirements are necessary to avoid a process that appears to be based on actual savings but might be highly inaccurate, biased, or random.

ASHRAE Guideline 14 recognizes three forms of M&V: whole facility, retrofit isolation, and calibrated simulation and includes compliance pathways for each approach. Guideline 14 does not have an equivalent for "Option A." But it does provide for short-term measurements under some circumstances, namely when both the baseline and reporting period loads can be considered constant. Table 6.2-1 "Retrofit Isolation Applications and Metering Required to Calculate Energy and Demand Savings" contains a detailed list of Retrofit Isolation situations and "required" metering.

Unfortunately, the guideline uses the term "measurement of savings" repeatedly, contributing to the potential for misunderstanding described earlier in this book in the section on *Concepts and Contexts.* It also does not delve into issues of contractual risk – the topic being considered beyond the technical guideline's purview.

Finally, ASHRAE Guideline 14 does not provide a specific solution for handling non-routine adjustments, and significant changes in factors affecting energy use, not typically accounted for in the baseline or reporting period. These adjustments require a unique, case-by-case analysis for accurate resolution due to their complex nature.

Inset: Excerpts from ASHRAE Guideline 14: Section 4.4.4

> Every project should find its own balance between the benefits and costs of measurement and resultant accuracy. Any of the paths can be implemented to suit a range of costs and certainties. Users are cautioned that even two well-experienced modelers will not generally determine the same savings amount, sometimes with significant differences.
>
> . . .
>
> Because of the trade-off between cost and uncertainty, the optimal approach for a specific project usually results from an iterative approach, where incremental improvements in accuracy are assessed relative to the increase in measurement cost. Such optimization requires that a value be placed on the level of accuracy. One way to accomplish this is to consider the uncertainty of the proposed approach by calculating results using the highest and lowest values in the confidence interval. The difference between these values can be translated into a dollar amount that is at risk. It can then be determined whether further expense for improving the savings measurement process is warranted to reduce this uncertainty.

3.3.4 ISO 50006 and 50015

Recall that ISO has published a suite of guidance documents relevant to the work of M&V, including ISO 50006 and 50015.

ISO 50006 deals with conducting energy performance assessments and energy performance measurements, including measurement and verification. It addresses the issue of measuring the impacts of energy management activities by emphasizing the importance of using appropriate measurement techniques and establishing clear and consistent baseline and reporting periods, while acknowledging that some energy performance improvements may not be easily quantifiable. It also provides guidance on how to determine the uncertainty of measurements and how to report results.

ISO 50015 focuses on the planning, conducting, reporting, and improvement of energy performance. The standard outlines the general principles of M&V, including the development of an M&V plan, the selection of data collection methods, and the analysis of data. It includes specific requirements for the determination of energy baselines and energy performance indicators, as well as guidance on measurement methods, including metering and monitoring. It also covers the use of M&V in evaluating the performance of energy-saving measures, such as energy management systems and renewable energy systems. Finally, it provides guidance on the use of uncertainty analysis in M&V, including the quantification of measurement errors and uncertainties.

EnPI

EnPI (energy performance indicators) are commonly used in the context of energy management and benchmarking. ENPIs are used to measure the energy performance of a facility or system.

In M&V, ENPIs can be used to establish baselines, set energy performance targets, and track progress toward meeting those targets. By monitoring and analyzing energy consumption data using these indicators, organizations can identify areas of inefficiency and implement energy-saving measures to improve performance and reduce costs.

ENPIs can be used in conjunction with M&V protocols such as IPMVP and ASHRAE 14 to provide a more comprehensive understanding of energy performance and impacts achieved through energy efficiency measures.

3.3.5 Others

Numerous other guidelines, protocols, and standards pertaining to building energy design, performance, and impact assessment may be relevant depending on your context.

- **LEED:** Leadership in Energy and Environmental Design is a rating system for the design, construction, operation, and maintenance of green buildings.
- **BREEAM:** Building Research Establishment Environmental Assessment Method is a standard for sustainable building design and construction.
- **CEN/CENELEC:** The European Committee for Standardization and the European Committee for Electrotechnical Standardization have developed standards for energy performance assessment of buildings.
- **Saudi Arabia M&V Protocol:** As part of its Saudi Energy Efficiency programs, Saudi Arabia has introduced protocols for measurement and verification to foster energy efficiency and sustainability in the country.
- **Dubai M&V Protocol:** This protocol was implemented by the Dubai Supreme Council of Energy to evaluate and report the impact of energy management activities in buildings and other structures as part of Dubai's energy conservation strategy.
- **Abu Dhabi M&V Protocol**[4]
- **Australia's Best Practice Guide to Measurement and Verification of Energy Savings:** Developed by the Energy Efficiency Council of Australia, this guide outlines best practices in designing, implementing, and reviewing M&V plans to ensure effective energy management activities.
- **The World Bank's Assessing and Measuring the Performance of Energy Efficiency Projects:** This report provides a framework for measuring and assessing the performance of energy efficiency projects, allowing for increased transparency and accountability in international development projects.
- **USAID's Introduction to Impact Evaluation Design: Planning and Managing Impact Evaluation:** This guide provides a step-by-step framework for the design, planning, and management of impact evaluations, with a particular emphasis on energy-related initiatives. It outlines methodologies to evaluate the impacts of energy management activities,

[4] https://www.doe.gov.ae/-/media/Project/DOE/Department-Of-Energy/Media-Center-Publications/Guidance-document/Abu-Dhabi-M-and-V-Protocol.pdf

ensuring the efficacy of USAID-supported energy efficiency programs in diverse global contexts.

* * *

In Chapter 2, we provided an overview of the fundamental concepts of M&V and emphasized the importance of understanding the context within which they are to be applied. In this chapter, we explained how standard guidance documents attempt to package the core concepts into language that can be adopted in contracts and program rules.

In the next chapter, we evaluate the moving parts of the M&V engine. We illuminate some of the nuances faced in applying M&V and point to sources of more information.

4

Practical Considerations

4.1 M&V Practical Considerations – Introduction

M&V is a fundamentally simple process in theory but potentially very complex in application. As an M&V professional, you must ensure that all stakeholders in the M&V process understand and agree on how M&V will be implemented in the context of their project. As we have seen in previous chapters, M&V relies on both measurements and models. It applies the counterfactual method, implying that there cannot be a direct measurement of "savings." M&V implementation incurs costs, and these costs must be justified through some approach to assigning value to the benefit of conducting M&V. All these realities – and decades of use of the current protocols – make clear that the most critical thing an M&V professional can do is to work with stakeholders to create a reasonable plan, maintain good communication, and manage expectations throughout the process.

There are many highly technical terms and areas where specialists must be brought in. Their terminology needs to be vetted and agreed upon throughout the process. And whenever possible, try to identify and work through potential issues before engaging in the process.

This section details specific issues to be addressed in the M&V *plan* and how they can *play* out in practice to arrive at a harmonious *settlement*.

* * *

"Plan, play, and settle" is a structured approach to M&V. The process is like learning to play a game: you learn the rules, devise a strategy, implement it, evaluate the outcome, and adjust accordingly – and repeat.

During the "plan" phase, the groundwork for the M&V process is laid out, and things like the project's scope, objectives, baseline for comparisons, and key performance indicators (KPIs) are defined. In this phase, you also draft an M&V plan that details how data will be collected, analyzed, and reported, putting all stakeholders on the same page regarding the methodologies and procedures to be employed.

"Play" refers to the active implementation of the plan. Here, you and your team collect and analyze data as outlined in the M&V plan. One key component of this phase is communicating the results to all stakeholders, including project sponsors, investors, and regulatory bodies.

Finally, the "settle" phase includes the verification of the project's results and the resolution of any lingering issues. But it is not just about concluding the project – it is also an opportunity for feedback and reflection.

This "plan, play, and settle" approach helps guarantee a structured, systematic, and iterative M&V process, where the results are reliable and serve as the basis for continuous improvement across subsequent projects.

4.1.1 Implementing the counterfactual: MM&V

The counterfactual method is a fundamental concept in M&V because it is the basis for the determination of impacts (energy "savings") achieved by an energy efficiency project. The method requires establishing and adjusting a baseline model of energy use for a given facility, which represents what the energy consumption would have been in the absence of the energy efficiency measures. Once the baseline model is established, the actual energy consumption is measured and compared to the adjusted baseline model to determine the impact.

There are a wide range of models that can be utilized to implement the counterfactual method. Identifying the appropriate model type and specific parameters is a fundamental task for the M&V professional.

The two main types of models used in M&V are statistical models and engineering models.

Statistical models use historical data and statistical algorithms to document relationships of variables that drive energy use. Engineering models, on the other hand, are based on the physical principles of the system being measured and can provide a highly detailed analysis of energy usage. We will delve into both types of models in more detail in Section 4.2.

4.1.2 Causality first[1]

M&V is not unique in the use of models and the counterfactual method to assess the impact of interventions. In many areas of business and science,

[1] https://en.wikipedia.org/wiki/Causal_inference

similar methods are applied. A common warning in these more general areas is the need to ensure that the models (and algorithms) used in any analysis are based on actual physical relationships – i.e., any model should first be assessed as "causal." Fortunately, for most applications in M&V, the causal model assumption is straightforward. If the facility has a heating/cooling system controlled by a thermostat, then outdoor temperature will impact energy use and the physical relationship is valid. Most building energy simulation model software packages are based on validated scientific and engineering algorithms.

An M&V professional can ensure that the model used to implement the counterfactual model is causal by following the below steps:

- Develop a clear understanding of the underlying physics and engineering principles that govern the system being analyzed. This includes identifying the key variables and how they interact with one another.
- Use this understanding to develop a conceptual model of the system that captures the key cause–effect relationships.
- Translate this conceptual model into a mathematical model that can be used to simulate the behavior of the system over time.
- Validate the model using real-world data and adjust it as needed to improve its accuracy.
- Use the model to conduct sensitivity analysis to determine how changes in key inputs affect the outputs.
- Test the model's predictive accuracy by comparing its outputs to actual data from the system.
- Use the model to conduct counterfactual analysis, carefully selecting the appropriate comparison group and controlling for any confounding variables.

4.1.3 Longitudinal and cross-sectional

Before going into more detail on the most common types of models, remember that counterfactual analysis can be applied in two ways: longitudinally and cross-sectionally.

The primary focus of this book is longitudinal. Longitudinal counterfactual analysis, as previously mentioned, is performed by comparing the energy use of a single facility before and after an intervention.

Cross-sectional counterfactual analysis, on the other hand, compares the performance of a facility before and after an intervention, such as an energy efficiency upgrade, with the performance of a group of similar buildings that did not undergo the intervention. This allows for a comparison of the energy consumption and cost impacts with the assumption that any differences in performance can be attributed to the intervention. Cross-sectional analysis is related to "comparison groups" and randomized control trials (RCT). In some branches of science, this is considered the "gold standard."

Both longitudinal and cross-sectional counterfactual analyses are commonly used in the evaluation of demand-side management (DSM) programs.

While both types of counterfactual analysis can be used in M&V to estimate the impacts of an energy efficiency project or program, for the purposes of this book, we will focus on longitudinal analysis.

4.1.4 Terminology

"M&V" is the application of the counterfactual method to energy management projects. As we saw previously, "M&V" can be thought of as either a general concept or the implementation of guidance documents. Each of these documents has adopted slightly different terminology, and each has a different approach to taxonomy and categorization.

What follows is terminology common to all these documents, but perhaps with a new interpretation. As always, the most important consideration for the M&V professional is clear and consistent communication with stakeholders about these general concepts, leading to successful settlement.

4.1.5 Pre/post: different from counterfactual

First let us expose one potential issue with terminology – the difference between "pre/post" and "counterfactual." When working with stakeholders, it is natural for some to interpret the concept of M&V as comparing an absolute amount of energy measured in the baseline period to the actual energy consumed in the reporting period. This is often called a "pre/post" or "before/after" comparison. For many people, this seems like the obvious way to calculate impacts. But the implementation of the counterfactual method requires the concepts of models and adjustments. It is best to ensure that everyone understands the concepts of modeling and adjustments early in the M&V process.

Baseline

The baseline is a reference dataset collected during the baseline period. This data typically includes all the information required to build a baseline model, which can be adjusted in the reporting period.

The baseline contains the historical energy use data for a specific period. This is true, but there is much more information that needs to be captured in the baseline period to fully document the baseline and support the development of the M&V plan.

To create a baseline for M&V, you typically need the following information:

- **Facility or system characteristics:** Information about the facility or system being evaluated, such as its size, layout, and operating characteristics.
- **Historical energy use data:** Data obtained from utility bills, submetering, or other sources.
- **Weather data:** Used to normalize energy use data and account for variations in weather conditions over time.
- **Occupancy and use data:** Information about how the facility or system is used, such as occupancy schedules, equipment schedules, and other relevant data.
- **Equipment specifications:** Detailed specifications of the building's equipment and systems, such as HVAC systems, lighting systems, and other energy-using systems.
- **Other relevant factors:** Other relevant factors that may affect energy use, such as changes in building occupancy, construction, or renovation activities or changes in energy pricing or utility rates.

Much of this data is typically collected during the energy audit,[2] a detailed assessment of a building's energy consumption that identifies areas of energy waste and recommendations for energy-efficient improvements. The audit may involve collecting information on building systems, equipment, and operational characteristics, as well as gathering energy consumption data for a given period. Practically speaking, as an M&V professional, you should be aware of any data already collected as part of the larger energy management process.

Finally, keep in mind one of the primary goals of the M&V professional: to ensure that all parties are aware of the tradeoffs and compromises required to create a "fair and relevant" baseline model for calculating the impacts of energy management activities.

[2]Handbook of Energy Audits, https://elibrary.riverpublishers.com/ebook_details.php?id=864

Baseline Period

The baseline period is the period of time selected to document the baseline conditions and create a baseline model. The "correct" baseline period is open to judgment and depends on the project's specific circumstances. It should be long enough to establish a representative pattern of energy consumption but not so long that it includes changes in building use, occupancy, or other factors not expected to be repeated in the reporting period. Remember to consider factors such as weather patterns, changes in building use or occupancy, and other relevant factors, such as the expected future use patterns.

Baseline Model

For most M&V plans, the baseline energy use data is analyzed, and a model is created to characterize the baseline performance. The baseline model is adjusted and compared to actual energy use to determine impacts.

When creating a baseline model for M&V, there are several major issues to consider, including:

- **Data availability and accuracy:** The baseline model relies on accurate and comprehensive data about the building and its systems, as well as historical energy usage data. The accuracy and completeness of this data will have a significant impact on the accuracy of the baseline model.
- **Representativeness of the data:** The data used to create the baseline model should be representative of typical building operation.
- **Normalization and weather adjustment:** Energy usage can vary significantly due to factors such as weather, which can make it difficult to compare energy usage over time. Normalization and weather adjustment techniques can help to account for these factors, allowing for more accurate comparisons of energy usage.
- **Model complexity:** Baseline models can range in complexity from simple regression models to more complex physics-based models. The level of complexity will depend on the available data and the desired level of accuracy – recall that there are often tradeoffs between level of certainty and cost.
- **Sensitivity analysis:** Conducting a sensitivity analysis allows you to identify key drivers of energy usage and the potential sources of uncertainty in the baseline model, which can point out areas where additional data or modeling may be needed.

When creating a baseline model, the M&V professional can choose between two types of models (statistical and physical – Section 4.2) and two system boundaries (whole building and retrofit isolation – Section 4.4).

Adjusted Baseline

The adjusted baseline is used in the implementation of the counterfactual method in M&V. It is created by adjusting (updating) the baseline model to reflect changes in factors that affect energy consumption but are *not* the result of the energy management activity.

The adjustment process – which should always be described in detail in advance in the M&V plan – involves the use of statistical analysis and modeling techniques to isolate the effects of the influencing factors and estimate the energy use that would have occurred in the absence of these factors.

Note that this definition of an adjusted baseline precludes any interpretation that the "baseline" data is changed. And the term "adjusting the baseline model" has two potential meanings: routine adjustments and non-routine adjustments.

Routine Adjustments

Routine adjustments are used to calculate adjusted baseline energy using *independent variables* obtained during the reporting period. This is also called "running the model" or "updating the model" to reporting period conditions. Statistical models will typically have one independent variable but may have more. Physical models will often have several variables or "dynamic factors" that can be used to update the model.

Non-routine Adjustments

Any change in the facility that occurs during the reporting period that was not explicitly parametrized (accounted for and factored in as a parameter) in the baseline model is called a non-routine event (NRE).

Non-routine events can include various factors such as changes in operations, added or removed loads, natural disasters, or other unforeseen circumstances that have a significant impact on the facility's energy use.

Depending on the measurement boundary, you can reasonably expect that many changes will occur in the reporting period. As an M&V professional, you *must make decisions about when to* adjust for non-routine events.

Static Factors

The fundamental theory of M&V is that practitioners can build and update (adjust) a baseline model throughout the reporting period. Much of the work of an M&V professional is to develop an acceptable "Fair and Relevant" baseline model. There are statistical tests for assessing the independent variables in the baseline model. However, all the other data collected as part of the baseline may also be subject to adjustment. This is a practical requirement that differentiates M&V from "science." In M&V, we accept that we cannot plan for every change to the baseline that may happen in the reporting period. The term of art that applies to the baseline data that is not used in routine adjustments is "static factors" (as opposed to the dynamic factors, aka independent variables).

4.2 Statistical (Inverse) and Physical (Forward) Models

Now that we can see how models are used in implementing the counterfactual method for M&V (or MM&V), let us look closer at the types of models that are used. For our purposes, we can define two categories: *statistical* and *physical.*[3]

Note: M&V models are all "time-series" models.[4]

Statistical (regression) models and physical (simulation) models are both used to represent and analyze baseline characteristics, but they have some key differences in terms of their underlying assumptions, methods, and applications.

Statistical models are based on actual energy data collected and analyzed in the baseline period. The goal of a statistical model is to establish the parameters (often slope and intercept of a line) that represent the behavior of the energy consuming system in the baseline period. This is also often called regression analysis.

Physical models, on the other hand, are used to represent the underlying processes and interactions that give rise to the energy data and are based on the idea of using mathematical or computational techniques to represent these processes. They often make assumptions about the rules and mechanisms that govern the system and use computational methods to simulate the behavior of the system over time.

Common physical models in M&V include models of individual systems, such as HVAC systems, lighting systems, and building envelopes, as well as whole facility models that combine the performance of all systems within a building. These models use physical principles, such as thermodynamics,

---

[3]The terms "Inverse andForward" are often used in lieu of "Statistical and Physical" in engineering literature.

[4]https://www.kaggle.com/learn/time-series

fluid mechanics, and heat transfer, to simulate the performance of a facility and its systems.

In general, statistical models are more concerned with estimating the relationships between variables using data, while physical/simulation models require an understanding of the underlying mechanisms and processes that give rise to the data. Both types of models have their own strengths and weaknesses, and the choice of which one to use will depend on the specific research question and the available data.

This classification of statistical and physical models is not unique to energy management and M&V. In *The Signal and the Noise*, Nate Silver discusses the challenges facing a wide range of disciplines that utilize models to attempt to predict the operation of complex systems. Indeed, prediction is a core concept of M&V, but we also use the models for comparison. As Silver notes, in most cases, a physical model is preferred – but this type of model is more difficult to create and update.

4.2.1 Statistical models – details

Statistical models, including regression, have been used in a variety of fields for over a century, and their use has evolved over time as new methods and techniques have been developed. Regression models were among the earliest tools used in econometric analysis, and they continue to be widely used today in various forms. In other fields, the use of statistical models and regression analysis varies depending on the discipline and the specific research questions being investigated. In the social sciences, for example, regression models have been used extensively in fields such as sociology, psychology, and political science.

The use of regression models in M&V dates to at least the 1980s, when researchers began using simple linear regression models to estimate baseline energy consumption and calculate the impact of energy management activities. These models typically used monthly utility billing data.

Inset: PRISM (Princeton Scorekeeping Methods: Peak Reduction Analysis Tool for Residential Energy)

This tool was developed in the 1980s as one of the first statistical models for M&V. It used regression analysis to estimate the energy impacts of weatherization measures in residential buildings. PRISM was widely used by utilities and energy efficiency programs throughout the 1990s.

A relatively new development is the deployment of advanced metering infrastructure (AMI), which has led to new possibilities for statistical modeling in M&V. AMI provides interval data, which is data that is collected at regular intervals, such as every 15 or 30 minutes, on energy consumption. This enables the development of more detailed statistical models of energy consumption patterns. By using interval data, statistical models can account for the variability of energy consumption over shorter time periods and capture the effects of short-term events, such as peak energy usage during certain times of the day or week. This can lead to more precise estimation of baseline energy consumption and more accurate calculation of energy impacts.

Independent Variables (Drivers of Variability)

In all cases, M&V exists because facilities have variability in their energy use, and M&V helps to rationalize this variability. Statistical models utilize independent variables to explain variability in energy consumption.

Independent variables are factors that may influence energy consumption but are not directly related to the energy-saving measures implemented. These variables can include weather conditions, occupancy levels, production output, and equipment usage. By controlling for these variables, you can isolate and calculate the impact of the energy-saving measures more accurately.

To identify the best independent variables to use, you should start by reviewing historical data and conducting site assessments to determine which variables are relevant to the facility. You should also consult with stakeholders, such as facility managers and operations staff, to gain insights into the facility's operations and potential sources of variability.

Once you have identified the relevant independent variables, you can decide whether to collect data on them throughout the measurement period. While it may seem desirable to identify all the variables contributing to variability, a core function of the M&V professional is deciding which variables to include in the baseline model.

Some typical independent variables used in M&V include:

- Weather conditions (temperature, humidity, and wind speed)
- Production output or occupancy levels
- Time of day or day of the week
- Equipment schedules and operating hours
- Lighting schedules and occupancy sensors
- HVAC set points and controls
- Building envelope characteristics (insulation, windows, etc.)
- Changes in operational procedures or maintenance practices

The specific independent variables used will vary depending on the type of facility, the energy systems in use, and the goals of the M&V project or program.

Weather Data

Weather data, specifically outdoor air temperature data, is the single most important independent variable in most M&V projects. Average temperature and heating/cooling degree days (HDD/CDD) are two different weather variables that can be used in energy analysis to account for weather-related impacts on energy consumption.

Average temperature is a measure of the overall temperature conditions for a given period, such as a day, week, or month. HDD and CDD are commonly used as weather proxies in energy analysis and M&V. HDD stands for "heating degree days" and CDD stands for "cooling degree days." They are both measures of how much the temperature deviates from a certain reference temperature over a period of time.

HDD is calculated by subtracting the average daily temperature from the reference temperature (usually around 65 °F or 18 °C) and summing the negative values over a period of time (typically a day, week, month, or year). This gives an indication of how much heating is required to maintain a comfortable indoor temperature.

CDD is calculated in a similar way, but using a reference temperature for cooling (usually around 75 °F or 24 °C). It measures how much cooling is required to maintain a comfortable indoor temperature.

Both HDD and CDD are used as proxies for energy use because they are correlated with the amount of energy needed for heating and cooling. For example, a higher number of HDD indicates that more heating is needed, which typically means higher energy use. Similarly, a higher number of CDD indicates that more cooling is needed, which also typically means higher energy use.

TMY (typical meteorological year) data is a dataset of hourly weather data that is used to simulate long-term weather conditions for a specific location. It is generated from a 30-year period of historical weather data and is used to represent the typical weather conditions for that location. TMY data is commonly used in energy analysis and M&V to estimate energy use over a long period of time (e.g., a year) based on weather data, as it provides a representative sample of the range of weather conditions that a building might experience during that time period.

Strengths and Weaknesses of Linear Models

For many people, "statistical model" equates to "linear model." Linear models are often perceived as the best fit for data because of their simplicity and ease of interpretation. In a linear model, the relationship between the dependent variable and each independent variable is assumed to be linear, meaning that the change in the dependent variable is proportional to the change in each independent variable.

This assumption is intuitive and easy to understand, making it an attractive option. Additionally, linear models are often the default option in many statistical software packages and tools, further contributing to their widespread use.

However, linear models are not always the best fit for data. In many cases, the relationship between the dependent variable and the independent variables may be nonlinear, requiring more complex modeling techniques. In addition, linear models may not capture important interactions or nonlinear relationships between variables, leading to biased or inaccurate results. Therefore, you need to carefully evaluate the appropriateness of linear models for a given dataset and problem and consider alternative modeling approaches if necessary.

Inset: The Inverse Modeling Toolkit (IMT)

ASHRAE's Project 1050 produced the Inverse Modeling Toolkit (IMT), which assists in the development of regression models of energy utilization in facilities. The software is capable of identifying both single-variable and multi-variable least-squares regression models as well as variable-base degree-day models and single-variable and multi-variable change-point models.

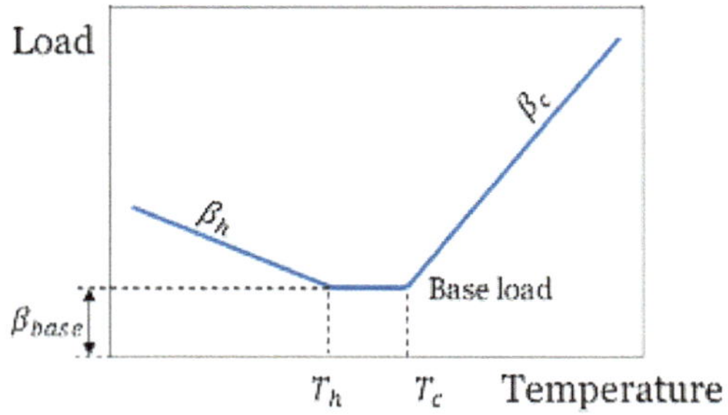

Figure 4.1 (Continued)

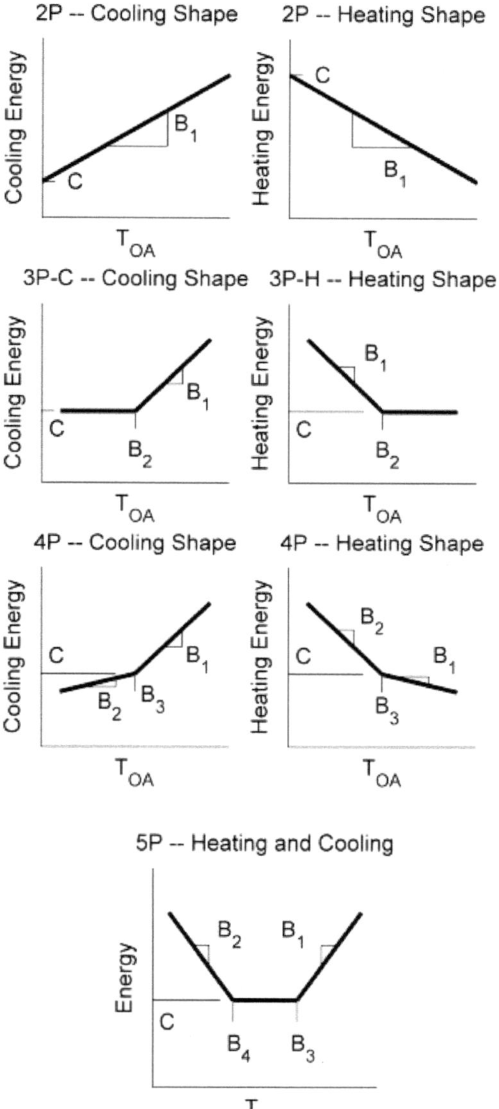

Figure 4.1 Common linear model forms.[5]

[5](**Energy Management Handbook 9th Edition,** https://elibrary.riverpublishers.com/ebook_read.php?id=828--Page783).

Inset: LBNL Time-of-Week and Temperature Model

An advanced type of change point model is the LBNL Time-of-Week and Temperature (TOWT) model, a regression model that is used to predict the hourly electric load of commercial and industrial buildings. The model was developed by researchers at Lawrence Berkeley National Laboratory (LBNL) and is based on data from over 1000 buildings in California.

The model uses two independent variables to predict hourly electric load:

- **Time of week:** The model includes 168 time-of-week indicators, one for each hour of the week. This allows the model to capture the fact that electric load varies by day of the week. For example, electric load is typically higher on weekdays than on weekends.
- **Temperature:** The model includes seven temperature bins, one for each range of outdoor air temperatures. This allows the model to capture the fact that electric load varies with outdoor air temperature. For example, electric load is typically higher on hot days than on cold days.

The TOWT model has been shown to be accurate in predicting hourly electric load for commercial and industrial buildings. It has been used by a variety of organizations, including utilities, energy efficiency program administrators, and facility owners.[6]

Much more information on building statistical models can be found in the Bonneville Power Administration's "Regression for M&V: Reference Guide."[7]

As mentioned earlier, the product of statistical analysis is typically the derivation of a limited number of parameters that will be used in the baseline model. These parameters represent a "best fit" of the statistical model to the baseline data. In all the M&V protocols mentioned above, the statistical approach to deriving those parameters is based on "Frequentist" methods. Many consider this the "standard" approach. But in recent years, there has been an increase in interest in an older, but in some ways more sophisticated statistical approach called Bayesian.

Frequentist statistics is a traditional statistical approach that focuses on estimating the probability of obtaining a certain result under repeated random sampling.

Bayesian statistics, by contrast, is an approach that treats probability as a measure of belief or uncertainty, rather than as a frequency of occurrence.

---

[6] https://buildings.lbl.gov/emis/assessment-automated-mv-methods

[7] https://www.bpa.gov/-/media/Aep/energy-efficiency/measurement-verification/3-bpa-mv-regression-reference-guide.pdf

Bayesian models in M&V use a prior probability distribution, which represents existing knowledge or assumptions about the variables in the model, and update this distribution based on observed data to estimate the posterior probability distribution. Bayesian models are useful when there is limited data available or when prior knowledge is important in making inferences.

Both Frequentist and Bayesian methods can be used to build statistical models for M&V, but each approach has its strengths and weaknesses. Frequentist methods are more commonly used in M&V and are easier to interpret, but they can be less flexible in handling complex relationships between variables. Bayesian methods are more flexible and can incorporate prior knowledge, but they can be more computationally intensive and require more expertise to interpret the results.

Bayesian methods, which have gained popularity as computing power has become more readily available, can also be used to generate regression coefficients and estimate uncertainty. Another advantage is that Bayesian methods can handle complex models and datasets. See Section 4.3 for more information.

4.2.2 Physical models – details

Physical models are mathematical representations of the physical processes that determine energy consumption and production in a facility or system. In M&V, physical models range from sophisticated energy simulation tools to simple spreadsheet models.

Generally speaking, simulation modeling could be possible for any energy-consuming system. But in the context of practical M&V, energy simulation models typically refer to building simulation.

Simulations use physics and engineering principles to model the energy use of a facility under different conditions. These simulations are typically run using specialized software that considers factors such as the building's geometry, insulation, HVAC system, and occupancy patterns. By inputting different scenarios, such as changes to equipment or occupancy schedules, the energy impacts of changes can be estimated. IBPSA USA maintains an excellent website with a full range of resources supporting their mission "Fostering better building performance through simulation."[8]

The main difference between a simulation model and a simple physical model is the level of detail and complexity captured. Simple physical models

[8]https://www.ibpsa.us/about/

often use simplified equations or assumptions to describe the behavior of a system or process, while simulation models are typically more detailed and complex, accounting for a wider range of variables and interactions.

ASHRAE Standard 140: Standard Method of Test for the Evaluation of Building Energy Analysis Computer Programs is the ASHRAE standard that defines the minimum requirements for facility energy simulation software.[9] This standard provides a methodology for evaluating the accuracy and reliability of facility energy simulation software, including guidelines for testing software against a set of benchmark models. The standard also includes requirements for software documentation, model input and output formats, and model calibration procedures. By establishing a consistent methodology for evaluating building energy simulation software, ASHRAE Standard 140 helps ensure that simulation results are accurate and reliable. The development of Standard 140 is ongoing.

The advantage of spreadsheet models in M&V is that they offer a more transparent and easy-to-understand approach. They use basic equations and assumptions to estimate energy use, such as the amount of electricity used by lighting, motors, fans, pumps, or other systems. While not as detailed as simulations, spreadsheet models can often be adequate in many cases.[10]

Calibration of Models

The algorithms in the most popular simulation software have been tested using a variety of methods, including comparison to real-world data. However, models of actual project facilities also need to be calibrated to be useful for M&V purposes. Here, calibration involves adjusting the inputs of the simulation model to match the actual energy consumption of the facility in the baseline period. This process can be challenging, as there can be many factors that influence facility energy use such as plug loads, HVAC system efficiency, and occupancy schedules. The calibrated model can then be used as the adjusted baseline model.

The calibration process can be time-consuming, and it requires a significant amount of data. It also requires a skilled professional with expertise in facility energy modeling and M&V.

Here again, the latest generation of M&V researchers are finding ways to apply Bayesian analysis to physical models.[11]

---

[9] https://energy-models.com/blog/what-best-energy-modeling-software

[10] https://www.buildingsandcities.org/insights/commentaries/misuse-building-performance-simulation.html

[11] https://ideaslab.io

Other Uses for Physical Models

Energy simulation/physical models can be useful for other M&V purposes, such as non-routine event calculation and interactive effect calculations, because they can provide detailed and accurate predictions of energy consumption and performance.

Non-routine events, such as equipment failures or unexpected changes in occupancy, can significantly impact building energy consumption and performance. Energy simulation models can be used to estimate the energy impact of these events by modeling the system both before and after the event occurs.

Interactive effect calculations involve modeling the interactions among multiple energy efficiency measures – for example, the impact of installing a new HVAC system on lighting energy consumption or the impact of adding insulation on space heating and cooling. Energy simulation models can be used to model these interactions and estimate their impact on overall energy consumption and performance. This can help identify the most effective combinations of energy efficiency measures and inform decision-making related to building upgrades and retrofits.

4.2.3 Practical modeling

George Box's well-known quotation – "All models are wrong, but some are useful"[12] – emphasizes that while no model can perfectly capture the complexity of reality, models can still be valuable tools for making predictions and understanding relationships.

This acknowledges that simplifications are necessary to make models tractable. However, it also recognizes the importance of selecting the appropriate level of complexity for the problem at hand. In the context of M&V, this means that practitioners must carefully choose the appropriate model and level of detail to balance accuracy with cost, practicality, and feasibility.

For example, if a model assumes that a facility operates under steady-state conditions but the facility's operations are actually highly variable due to occupancy patterns, weather, or other factors, then the model will not likely

[12]https://www.lacan.upc.edu/admoreWeb/2018/05/all-models-are-wrong-but-some-are-useful-george-e-p-box/

provide accurate results, and a more complex (and more costly) model that accounts for these factors may be necessary.[13]

Inset: The Spherical Cow

The concept of the "spherical cow" refers to the practice of simplifying real-world problems for the sake of the feasibility of computational analysis – especially when the simplification undermines the model's practical applicability.

The phrase "spherical cow" originated in this joke, told in many ways over the years:

A dairy farm's milk production was low; so the farmer asked the local university for help. A multidisciplinary team of professors was assembled, headed by a theoretical physicist, and weeks of intensive on-site investigation took place. The scholars then returned to the university, notebooks filled with data, where the task of writing the report was left to the team leader. Shortly afterward, the physicist returned to the farm, saying to the farmer, "I have the solution – but it works only in the case of spherical cows in a vacuum."

Imperfect Models and Types of Uncertainty

As we dive deeper into the methods of M&V, we get a better idea of the challenges of implementing the counterfactual method in real-world situations. The baseline model needs to be as accurate as possible but also be relevant to the expected conditions of the reporting periods. And when, inevitably, changes occur in the reporting period that were not anticipated (parametrized) in the baseline model, "non-routine" adjustments must be made.

Another way of thinking about these challenges is called uncertainty analysis. Within uncertainty analysis, there are two main types of uncertainty: aleatory and epistemic.

- *Aleatory uncertainty* is due to the inherent randomness and variability in a system that cannot be reduced, even with more data, better modeling techniques, and additional financial resources.
- *Epistemic uncertainty* arises from our lack of knowledge or understanding of a system. This type of uncertainty *can* sometimes be reduced with the right tools and sufficient effort. But in many cases, events that were not foreseeable during the baseline must be addressed in the reporting period.

[13]Kaggle held a competition that repeated earlier efforts to identify the best models for forecasting the energy use of buildings; the project and results can be found at https://www.kaggle.com/c/ashrae-energy-prediction/overview/prior-competitions

When we apply M&V, we try to estimate the impact of a particular energy management measure or program. However, there is always some level of uncertainty in our results due to aleatory and epistemic uncertainty.

Non-routine events can introduce both types of uncertainty, as they may stem from external factors (aleatory) or be related to a lack of information about the system's behavior (epistemic).

We are not alone. The concepts of aleatory and epistemic uncertainty, as well as the need to consider model selection and model assumptions, are not unique to M&V. Indeed, they are relevant to many disciplines, including engineering, physics, economics, and finance. The terminology may vary, but the underlying concepts are the same. Whatever the field, the main idea is that it is important to understand and (attempt to) quantify sources of uncertainty and risk in any modeling or analysis process, to be able to make informed decisions about model selection and adjustments and properly interpret the results.

All Models are Wrong, All Buildings are Broken

Another challenge to implementing the counterfactual method through modeling is the fact that many facilities do not operate as designed. In other words, many of the buildings and other types of facilities or systems on which we perform M&V are "broken."

A facility's condition can change over time due to a variety of factors, such as renovations, equipment failure, and the steady creep of entropy. But modeling packages only makes provision for nominal system performance (there are no inputs for many common system failures). This is especially true with energy simulation models. Energy simulation models rely heavily on the accuracy of inputs such as building geometry, occupancy schedules, HVAC system performance, weather data, and equipment operation. Any deviation from the assumed inputs can result in significant differences in predicted energy consumption compared to actual energy consumption.

Model Selection Requires Judgment

In M&V – or any other area of science or engineering – there is no perfect model. Instead, the perpetual challenge has been to arrive at a "good enough" model. Obviously, what constitutes "good enough" will depend on needs, goals, and other elements of the context and stakeholders. Reasonableness on all sides is required.

When selecting or developing a model, you need to exercise professional judgment to determine which variables to include and how to specify the model. Interpreting a model's results requires professional judgment as well. As an M&V professional, you must assess the realism of the model, the quality of the data, and the relevance of the model to the real world. How accurately does the model reproduce observed phenomena? How sensitive is it to different assumptions and input variables? Can all stakeholders agree to be realistic about the strengths and weaknesses of any particular solution?

Problematizing Judgment

Often, when we exercise judgment, we use mental shortcuts such as heuristics and biases that can undermine our best efforts to engage in rational decision-making.

Heuristics are practical "rules of thumb" that allow us to arrive at judgments and decisions quickly and efficiently. They are useful in many situations, but they can also lead to systematic errors if used in situations where more information or a more complex decision-making process is needed. For example, an M&V professional might succumb to the availability heuristic by judging the probability of an energy-conservation measure's effectiveness based on easily recalled, similar past projects. This could lead to suboptimal decision-making if the current project has different conditions not present in the previous ones.

A related, sometimes overlapping, concept is that *biases* are predispositions to favor a given conclusion over other conclusions, often based on preference, emotion, or recent information. For example, an energy auditor may fall victim to *confirmation bias* by unconsciously giving more weight to findings that confirm their preexisting beliefs about the energy performance of a facility, while downplaying or overlooking information that contradicts these beliefs.

We can never fully eliminate heuristics and biases from our thinking. At most, we can limit their influence by recognizing the role that they play in our judgment and decision-making.

Inset: Finite Transcendence

Heuristics and biases connect to Plato's concept of "finite transcendence," which suggests that the human mind is limited in its ability to understand the true nature of reality. Plato argued that we can access only a limited, imperfect version of reality and that true understanding must be obtained through a process of transcendence that goes beyond the abilities of the human mind.

Data, Information, and Wisdom

Data, information, and wisdom can be thought of as a hierarchy:

- At the bottom of the hierarchy is *data*, which refers to raw, unprocessed facts or figures. Data can come in many forms, such as numbers, text, images, or audio. On its own, data is not particularly useful, but it has the potential to be transformed into something more meaningful.
- The next level of the hierarchy is *information* – data that has been organized, structured, and processed in a way that makes it meaningful and useful.
- The highest level is *wisdom*, the ability to use information to make sound judgments and good decisions in life. Wisdom is the ability to understand the big picture, weigh the long-term consequences of actions, and consider multiple perspectives – to use information and knowledge to navigate difficult situations and make informed and ethical decisions.

The Perils of Overfitting: Do Not Fit the Noise

Modeling *can* be taken too far, through a phenomenon known as overfitting.[14] An overfitted model is one that is too complex and fits the underlying data too closely, such that it captures noise or random variations in the data rather than underlying patterns or trends. This occurs when a model has too many parameters relative to the amount of data on which it was based. Such a model may perform well on data similar to what was used to develop it, but it will lack flexibility and lose predictive ability as circumstances change.

4.3 More on Bayesian Methods

A topic of increasing interest to the M&V community is the use of Bayesian statistical techniques.

Bayesian approaches use probability to represent uncertainty – both the uncertainty about the values of the model parameters and the uncertainty about the data. This allows for a more flexible and expressive model of uncertainty compared to traditional frequentist approaches, which only allow for uncertainty about the data. One way that Bayesian statistics can be useful when using the counterfactual method is by allowing for the incorporation of prior knowledge about the parameters of the model. In a counterfactual

[14] https://see.stanford.edu/Course/CS229/42

analysis, this can be used to incorporate information about the factors that are known to influence the outcome, even if there is limited data available.

Google's CausalImpact[15] is a good example of a tool for simple Bayesian analysis for M&V because it demonstrates how to use a Bayesian framework to estimate the impact of an energy management project.

The CausalImpact code uses a Bayesian structural time-series model to estimate what the outcome variable (e.g., energy consumption) would have been in the absence of the intervention and then compares it to the observed outcome variable to estimate the causal effect of the intervention. This approach is particularly useful for M&V because it allows for the incorporation of prior knowledge about the system and provides a probabilistic estimate of the causal impact, including uncertainty.

Furthermore, CausalImpact's code provides a practical example of how to implement a Bayesian analysis in R, which is a commonly used tool in M&V. By following the code and modifying it for their own purposes, M&V professionals can gain a better understanding of how to use Bayesian methods to analyze time-series data and estimate causal effects.

For users of Python, CausalPy[16] has similar capabilities.

John Shonder[17] and Herman Carstens[18] conducted original research on using Bayesian methods for M&V. Simon Rouchier has continued this work.[19]

A very good source of general information is: https://xcelab.net/rm/statistical-rethinking/

Another excellent source of information on Bayesian tool is: https://statmodeling.stat.columbia.edu/category/bayesian-statistics/

4.3.1 Stan

Stan is a programming language commonly employed for performing complex Bayesian analyses. It allows users to build models using a high-level syntax and then automatically generates optimized C++ code for running Bayesian inference.

[15]https://google.github.io/CausalImpact/CausalImpact.html

[16]https://github.com/pymc-labs/CausalPy

[17]https://www.researchgate.net/publication/268095106_Bayesian_Analysis_of_Savings_from_Retrofit_Projects_SA-12-003

[18]https://www.mdpi.com/1996-1073/11/2/380

[19]https://srouchier.github.io/tutorials/

One advantage of Stan is its ability to handle complex models with many parameters and data points. Users can also easily incorporate prior knowledge into models, which can be particularly useful in M&V, where prior knowledge about the facility or process being analyzed can help inform the analysis.

Another advantage of Stan is its efficiency in sampling from complex posterior distributions. It uses a hybrid Monte Carlo algorithm that can adapt to the shape of the posterior distribution, which can lead to faster and more accurate results compared to other sampling methods.

Here are some references for learning more about Stan:

- **Stan User's Guide:** This is the official user guide for Stan, which provides a comprehensive overview of the software and its features. It includes examples, tutorials, and a reference for the Stan language. You can find the user's guide at https://mc-stan.org/docs/stan-users-guide/index.html.
- **Stan Reference Manual:** This is the official reference manual for the Stan language, which provides detailed information about the syntax and semantics of the language. You can find the reference manual at https://mc-stan.org/docs/reference-manual/index.html.
- **Stan Forums:** This is an online community where users can ask questions, share knowledge, and collaborate on projects related to Stan. You can access the forums at https://discourse.mc-stan.org/.
- ***Bayesian Data Analysis*, Third Edition:** This is a book by Andrew Gelman, John B. Carlin, Hal S. Stern, David B. Dunson, Aki Vehtari, and Donald B. Rubin that provides a comprehensive introduction to Bayesian data analysis using Stan. It includes case studies and examples using real-world datasets. The book can be found at https://www.stat.columbia.edu/~gelman/book/ and is freely available in electronic form.
- **StanCon:** This is an annual conference focused on Stan and Bayesian data analysis. The conference includes talks, workshops, and tutorials on a variety of topics related to Stan. You can find the conference website at https://mc-stan.org/events/.

Researchers at the National University of Singapore developed specific instructions on using STAN for building model calibration.[20]

[20]https://github.com/adChong/bc-stan

4.4 System Boundary

The fundamental theory of M&V, the counterfactual method, can be applied at different levels of analysis, as defined by the *system boundary*. This boundary can apply to any system or process, such as a facility or an industrial manufacturing process. Boundary selection is important because there are tradeoffs when implementing at one level vs. another. One of the M&V professional's major roles is to select an optimal system boundary.

The purpose of the system boundary is to define the precise limits of the system being evaluated for performance and/or "savings." It establishes what is included in the scope of the M&V plan and what is not and allows us to identify potential interactions or interdependencies with other systems. The system boundary can be defined in various ways, depending on the project and the goals of the M&V plan, but it generally encompasses all the components necessary for the system to function as intended, such as the equipment, controls, and associated infrastructure.

The primary categories of system boundaries in M&V are:

- **Whole facility:** The whole facility, including all energy-using systems and equipment within the facility, is coterminous with the system boundary. The facility boundary is often the same as the area served by the utility meter. This is the most comprehensive system boundary and is typically used for complex retrofits with multiple ECMs or new construction projects.
- **System-level:** The system-level boundary includes a specific energy-using system, such as the HVAC system, lighting system, or process equipment. This boundary is often used for projects that involve upgrading or replacing specific systems.
- **Subsystem-level:** This boundary includes a part or component of an energy-using system, such as an air handling unit, a lighting fixture, or a pump. This boundary is used for projects that involve upgrading or replacing individual components within an energy-using system.

4.4.1 History and terminology

As discussed in Chapter 3, the FEMP and IPMVP "options" terminology arose from an awareness that M&V is needed to comprise a range of approaches to allow M&V professionals to develop an optimal plan for each project. We have now discussed three fundamental distinctions in "M&V":

- performance vs. impact ("savings"),

- statistical models vs. physical models, and now,
- whole facility vs. "retrofit isolation."

The original authors of the protocols recognized that each project may be handled differently depending on various factors that could be fully appreciated only after an on-site investigation. For example, a facility might already have submetering installed in some cases, allowing for detailed M&V on end-use systems. In other cases, such as many utility programs, metered data from installed smart meters (AMI) is used to execute a "whole facility" approach. Once again, the task for the M&V professional is to ensure that all stakeholders grasp the strengths and weaknesses of each approach.

4.4.2 Whole facility approaches

It looks easy, but it's not. – Hassan Younes

Whole facility M&V is based on a simple premise – that it is possible to quantify impacts of interventions using the energy consumption data for the whole facility. The fact that many facilities have a single meter serving each electrical and gas load makes this method even more enticing.

When demonstrated in Excel or on paper, it looks easy. But in practice, it is not.

The counterfactual in this case is a "spherical cow" model of the baseline. It is never possible to create a model with detail on every factor influencing energy use. The M&V professional uses their judgment to balance the cost of M&V with a plan that captures the most important factors influencing energy use.

The model can be either statistical or physical. Each has its advantages and disadvantages.

Whole facility methods (Table 4.1) have two main potential issues:

- First, are the expected impacts of sufficient magnitude to be quantified?
- Second, will there be significant changes to the facility in the reporting period that might require non-routine adjustments?

Other questions that must be answered for a particular project are:

- The "signal" from the whole facility meter may show "savings," but do you know what "caused" them? Do you care?
- Is it important to determine the performance of individual ECMs?

Example: An ESPC is signed, and the M&V method selected is FEMP/IPMVP Option C. "Savings" under this contract are determined by comparing the adjusted baseline model (statistical) to the actual consumption.

Table 4.1 M&V option terminology.

Approach	**IPMVP and FEMP**	**ASHRAE**
Statistical	Option C	Whole facility
Physical (simulation)	Option D	Calibrated simulation

Approach	**Strengths**	**Weaknesses**
Statistical	• Simple to understand • Data can be easily collected • Captures all interactive effects • Model can be built at many levels of granularity, providing insights into operations • Uses utility billing data • Amenable to FLEX Market	• Does not report measure level impacts • Non-routine events can be a challenge to deal with
Physical (simulation)	• Captures interactive effects • Can be used for non-routine adjustments • Can be used for rate/tariff analysis	• Expensive and difficult to build • Calibration can be challenging

Without additional verification of the performance of the ECMs or detailed non-routine event monitoring, "savings" may occur with no assurance that the ECMs were what contributed. Furthermore, if the M&V is based on monthly billing data, there may be no way to determine which ECMs contributed.

However, if the contract required FEMP/IPMVP Option D or a hybrid of Options C and A or B, stakeholders would likely be able to determine ECM-level results.

- Will the project require frequent non-routine adjustments?

Another dimension to the M&V plan is the time interval for the metered data. Monthly utility billing data is still commonly used. But hourly data is now becoming more common. In any case, each analysis period of the reporting period must include accounting for non-routine events. If non-routine events are common, requiring non-routine adjustments, the cost of M&V can become untenable.

- Is it important to understand the "savings load shape"?

In a growing number of cases, the utility tariff is based on time of use or even real-time rates. Here, it is important to know the impacts of the ECM on an hourly basis. Hourly load shapes can be used to demonstrate "savings" that were "predicted."

Example: In a project in Saudi Arabia, the project implementor proposed turning off outside air fans at night. The baseline for the evening hours was quite stable. Given the magnitude of the proposed ECM, the impact should have been clearly visible during the evening hours immediately after implementation.

Looking forward (see Section 4.2.3.2), flex markets now provide incentives for projects that either reduce or shift loads away from peak hours.

Whole Facility Statistical Models ("Option C")

Clearly, the job of an M&V professional would be much easier if every facility had one meter that served all its loads, and the data from that meter was readily available.

Indeed, historically, whole facility (statistical) methods relied on utility bills – and many still do. However, while this is sufficient in many cases, there is no guarantee that the utility bill actually maps to the system boundary. Parking garages, streetlights, and other ancillary loads may be on the utility meter. And there may be more than one meter that serves the facility. Caution is warranted.

For this reason, as an M&V professional, you must study the facility's energy delivery systems (electrical wiring service, gas, and others) to confirm that the data collected corresponds to the desired measurement boundary. It is common in M&V to install dedicated meters on the appropriate circuits to collect energy data for the appropriate system boundary, whether whole facility or isolated system.

Using the whole facility measurement boundary is common in both ESPC contracts and utility programs. But there is typically a big difference. ESPC contracts have many clauses relating to the responsibility of each party for the performance and operation of the project. Utility programs often do not. In other words, when it comes to non-routine events and adjustments, ESPCs include specific guidance, but utility programs are often designed without including a method for accounting for non-routine events (see Section 4.7.3).

With the proliferation of "smart" meters and automated metering infrastructure (AMI), many utilities now offer programs under the banner of "pay for performance." These programs typically do not require the rigorous documentation of baseline conditions that one would expect in a performance contract, including a complete documentation of "static factors."

Instead, they rely on agreed-upon algorithms (see CalTrack and Open EE Meter) and operate under the assumption that non-routine events will "cancel out" in a large enough portfolio. In addition, some programs use control groups (the second dimension of the counterfactual) to provide information on how participants compare to the population at large.[21]

Whole Facility – Energy/Simulation Models ("Option D")

First principal energy models can also be used to implement the whole facility method. As mentioned earlier, a properly calibrated simulation model is the gold standard for M&V. It is also almost always prohibitively expensive. This situation is changing, but it will be a while before AI and the Internet of Things (IoT) combine to enable accurate simulations of complex systems.

In the meantime, the "spherical cow" principle applies, and you will need to exercise judgment in implementing whole facility physical models.

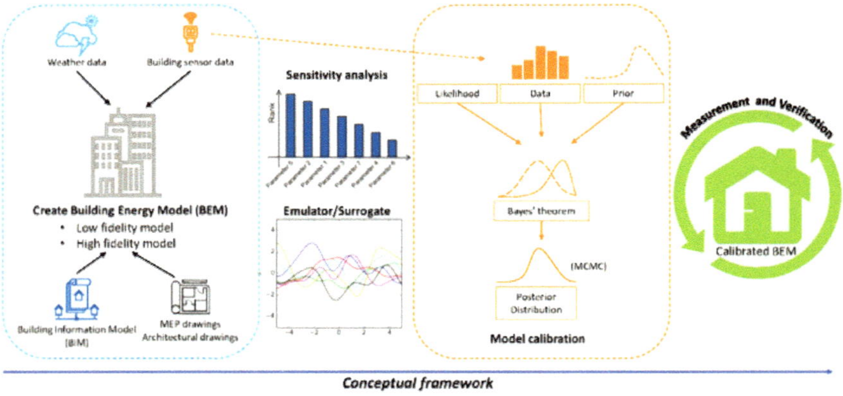

Figure 4.2 Conceptual framework for building energy model calibration.

A full description of energy modeling is beyond the scope of this book. The reader is encouraged to learn more by visiting the IPBSA site.[22]

Once again, given the level of skill and experience required to build competent energy models, it is unlikely that the typical M&V professional will be taking on this task. They will likely be designating it to a team member. However, it is well worth the effort to learn the basics of energy modeling to get a first-hand understanding of how they work. There is a wealth of training

[21] https://www.cpuc.ca.gov/-/media/cpuc-website/files/legacyfiles/n/6442463694-nmec-rulebook2-0.pdf

[22] https://www.ibpsa.us

available to at least become comfortable with the fundamentals and thus be able to interact with the experts.

New Construction vs. Existing

One circumstance where whole facility simulation models are frequently used is in new construction programs. In these cases, the baseline is hypothetical. Design information is used to create a simulation model of an entire facility. That model is later updated to reflect the implementation of additional energy management measures.[23,24]

4.4.3 System level (retrofit isolation)

Retrofit isolation is the term used when a system boundary is defined that only contains the equipment subject to the energy management project. *ASHRAE Guideline 14* defines the term as follows:

> An M&V approach that measures the energy consumption and demand of a portion of a facility or system, which has been upgraded or retrofitted, while keeping other parts of the facility or system isolated from the upgrade or retrofit measures.

The benefit of this approach is that the data collected will be much more likely to accurately reflect the predicted impacts. The major issue, however, is that there may be effects outside the system boundary that are not quantified. These "interactive" effects can occur both with the measured energy system (say, electrical power) and with other energy systems serving the facility (e.g., gas, steam, and chilled water).

Retrofit isolation methods introduce another set of complications as well. Primary among them is the need to install and maintain dedicated metering equipment. Accordingly, faced with the task of designing an M&V plan that includes installing (or utilizing existing) field metering to measure the energy consumption and significant parameters impacting it, you now must confront issues of cost, potential interruptions of facility operations, questions about the security of the M&V data, and safety concerns when installing and removing meters.

In later sections, we will cover all these issues in detail.

* * *

[23]https://www.sdge.com/sites/default/files/documents/2019%20SBD%20Handbook.pdf

[24]https://www.pge.com/en_US/large-business/save-energy-and-money/facility-improvement/new-construction.page

Imagine that you are tasked with determining the most effective approach to verify the impact of energy management activities for a building retrofit project. The facility owner is interested in maximizing impact while minimizing costs.

You can begin by conducting an assessment of the facility and identifying potential retrofit measures. You can then use modeling tools and historical energy data to estimate the quantified impact of each measure.

If the overall impacts are significant, you can recommend a whole facility approach – implementing multiple retrofit measures across the facility and then using a combination of monitoring and modeling to measure and verify the retrofits' total impact.

However, if the impacts are not significant enough to warrant a whole facility approach, you might recommend retrofit isolation. In this approach, targeted retrofit measures are implemented in specific areas of the facility where the potential for impact is the highest. You can then use monitoring and modeling to measure and verify energy management impact in those specific areas.

Ultimately, the choice between a whole facility approach and retrofit isolation will depend on a number of factors, including the facility's size and complexity, the available budget, the potential energy management impacts, and the level of accuracy required for M&V. You will need to carefully consider all these factors and work closely with the facility owner to determine the best approach.

* * *

4.4.4 Terminology

Performance and Operation

As discussed earlier, there is a distinction between two important concepts in characterizing energy-consuming systems: performance and operation.

This distinction can be useful for identifying the purpose of the M&V. Is it intended to verify that an ECM has been installed and is "performing" correctly? Or is it an attempt to quantify the impact of the ECM over an extended period? Both FEMP and IPMVP initially defined Option A as "verifying performance" and Option B as "continuous measurement in the reporting period." ASHRAE Guideline 14 does not have an "Option A," though it allows for short-term measurements for constant pre/post loads.

Referring to three fundamental distinctions in M&V approaches, verifying performance is directly related to intermittent measurements, whereas quantification of impact ("savings") requires continuous measurement:

- Intermittent (short-term) vs. continuous measurement during the reporting period
- Statistical vs. physical
- Whole facility vs. system/retrofit isolation

4.4.5 Hybrid systems

Defining a retrofit isolation for a complex system like compressed air or chilled water can be challenging because these systems typically have many interconnected components and subsystems that affect each other's energy use.

For example, in a compressed air system, the energy use of the compressor is directly related to the pressure and flow rate of the air it produces. However, the pressure and flow rate are also influenced by the distribution network of pipes and valves, which can leak or become inefficient over time. Additionally, the end-use equipment that the compressed air is serving (such as pneumatic tools or machinery) can vary in energy consumption depending on factors like workload, maintenance, and efficiency.

Similarly, in a chilled water system, the energy use of a chiller is affected by the temperature and flow rate of the chilled water, as well as the ambient temperature and humidity. However, these factors can also be influenced by the cooling towers, pumps, and heat exchangers that are used to regulate the temperature of the water. And like before, the energy use of the end-use equipment that the chilled water is serving (such as air handling units or process equipment) can vary depending on workload, maintenance, and efficiency.

Given these complexities, defining retrofit isolation for such systems can be difficult. One approach is to divide the system into smaller subsystems or components and analyze each one separately while accounting for their interactions with other parts of the system. It may also be necessary to incorporate sensors or metering at multiple points in the system to accurately capture the energy use of each component and its interaction with others.

4.4.6 Confounding situations and issues

There are many situations where the simple categories of whole facility vs. retrofit isolation are confounded. Here are a few examples:

- **Multiple meters:** Some facilities have multiple utility meters, each serving a different portion of the facility. This can make it difficult to determine the facility's overall energy consumption and isolate the impact of specific retrofit measures.
- **Parking garages:** In some facilities, parking garages are not included in the main facility energy metering system.
- **Large chiller plants that are partially metered:** Some facilities have large central chiller plants that serve multiple parts of the facility. These chiller plants may be only partially metered, which can make it difficult to determine the overall energy consumption of the plant and evaluate the impact of retrofit measures that target it.
- **Shared spaces:** In some facilities, spaces like lobbies, hallways, and stairwells are shared by multiple tenants or users.
- **Tenant improvements:** In some facilities, tenants may make improvements to their leased spaces that impact the overall energy consumption of the building. This can interfere with attempts to measure the impact of energy management measures by the facility owner or operator.

These are just a few examples of the complex and confounding situations that can arise in M&V. They highlight the importance of carefully considering the system boundary when designing M&V plans and working closely with all stakeholders to agree upon a suitable methodology.

4.4.7 Using utility meters is common – but not required

It is possible to install additional meters to capture more granular energy consumption data beyond what the utility meter provides. This can be useful for capturing data on specific systems or areas within a facility that may not be adequately captured by the utility meter. However, any additional meters must be calibrated and installed properly to ensure accurate data collection. Additionally, you may need to work with the utility company to approve any additional meters and remain in compliance with relevant regulations.

4.4.8 Dual fuel

When a facility is served by multiple fuel sources, it becomes challenging to determine how much energy is consumed by each fuel source and how they interact with each other. For example, if a facility is heated by both electricity and natural gas, it is not always clear how much energy is consumed by each fuel source, as they may interact with each other to heat the building.

4.4.9 Multiple systems

Another common issue is for a facility to have both a central chiller plant and individual air conditioning units. To address these issues, you might install additional meters or submeters to capture energy use from different systems or develop more complex models to account for the interactions between fuel sources.

4.5 Baseline: Additional Considerations

Previously, we defined the baseline in broad terms. We said that all the information collected during an agreed-upon period of operation is necessary to characterize and document the existing conditions and collect the data that might be needed in the reporting period for adjustments. In addition, we discussed the "denominator problem" – is the facility currently meeting all applicable codes and expected service levels?

If the existing conditions do not meet required or desirable service levels, but after the project is complete those conditions must be met, will all stakeholders agree that the baseline is fair and relevant? Reasonable people could disagree. And addressing this situation could either increase or decrease the calculated impacts.

Before documenting the baseline, there should be a thorough investigation of the facility's systems in order to identify any issues that need to be addressed. Ideally, once the issues have been resolved, the baseline can be documented based on the corrected performance of the building. In some cases, it may be necessary to adjust the baseline period to account for the period during which the building was not operating properly.

For example, the presence of burnt-out lamps during the baseline period could affect the baseline model. The lighting system's energy consumption in the baseline period would be lower than it should be, resulting in an artificially low baseline. Therefore, it is helpful to ensure that all equipment and systems are functioning correctly and in accordance with manufacturer specifications before establishing a baseline. If burnt-out lamps or other equipment issues are found, they may be fixed or replaced before the baseline is established, if stakeholders agree and program guidelines allow.

If the extent of operations in the baseline period varies, you should use your professional judgment to select the most representative operations level. What you select should be representative of typical operating conditions during the baseline period and consistent with the expected future conditions.

If there are specific future operating characteristics that are known or expected, they should be incorporated into the analysis through sensitivity analysis or adjustment factors, rather than being included directly in the baseline. This will allow for a more accurate representation of the potential impact of energy management activities, while maintaining the baseline period's integrity.

4.5.1 A valid baseline

A valid baseline for M&V purposes is a reference point against which the performance of a building or system can be compared to determine the impact of energy management activities. As mentioned above, the M&V professional should ensure that all stakeholders understand that the baseline operations will be the basis of comparison for however long the reporting period will be in effect. In other words, everyone must agree that the future operations will be similar to the baseline – or at least similar enough such that minor adjustments will allow for a fair comparison. In the case where major renovations are planned, there may be a limited relevance of the baseline consumption, and, hence, M&V may not be recommended.

* * *

4.5.2 What Characterizes a "Fair and Relevant" Baseline?

- **Consistency:** The baseline should be consistent with the historical energy consumption data, operational hours, and other factors that could affect energy use.
- **Normalization:** Baseline models should take into account factors that affect energy consumption but are outside the control of the project, such as weather, occupancy, and production levels.
- **Accuracy:** The baseline model should accurately predict energy consumption under normal operating conditions.
- **Transparency:** The baseline model should be transparent and easily understood by all stakeholders, including the project team and any third-party evaluators.

- **Repeatability:** The baseline model should produce consistent results when used by different stakeholders.
- **Simplicity:** The baseline model should be simple enough to be implemented and understood by all stakeholders.
- **Sensitivity analysis:** The baseline model should be evaluated using sensitivity analysis to assess the impact of different input parameters on the predicted energy consumption.

* * *

4.6 Special Baseline Considerations for Utility Programs

As an M&V professional, you need to ensure that the results of M&V meet the needs of the context. There is *not* only one definition of "savings" – or, as I have preferred to call it, impact. Impact can be defined in many ways. For example, while most energy-savings performance contracts use "existing conditions" for the baseline, many utility programs apply a different framework. For example, some utility programs use a "code baseline" rather than an "existing conditions" baseline. A code baseline represents the minimum efficiency standards that should have been in place at the time of implementation. This is to protect the interests of utility ratepayers so that they are not subsidizing projects where the facility owner should have already installed more efficient equipment.

The code baseline is based on established building codes, equipment standards, and other industry best practices for energy efficiency. The goal is to determine the minimum efficiency level that should have been in place at the time of implementation and then to quantify impacts of the energy efficiency measures against this baseline.

Code baselines sometimes cause confusion because they represent a theoretical minimum efficiency standard rather than an actual measure of energy consumption before implementation. To put it differently, the "savings" calculated using a code baseline may not accurately reflect the observed energy impacts of the energy efficiency measures.

For example, if a facility's energy consumption failed to comply with minimum efficiency standards at the time of implementation (for example, if they were still using incandescent lamps when LEDs are required), the "savings" calculated using a code baseline might be lower than the observed impacts of the energy efficiency measure.

A code baseline can also make it challenging to compare the energy management activity impacts across different facilities or programs because the code baseline may vary depending on the location, facility type, and other factors. Thus, while code baselines can help protect ratepayers, they often differ from the observed impacts of energy efficiency measures and may complicate comparisons across different facilities or programs.

4.6.1 Above code

"Above code" concepts refer to energy efficiency measures that go beyond the minimum requirements set by building codes and standards. These measures are often adopted voluntarily by facility owners and developers who wish to reduce long-terms costs, take advantage of incentives, or demonstrate their commitment to sustainability. Some building programs and rating systems, such as the U.S. Green Building Council's Leadership in Energy and Environmental Design (LEED) certification, offer incentives and recognition for buildings that incorporate "above code" measures.

Some examples of "above code" measures include:

- Building envelope improvements, such as increased insulation, high-performance windows, and air sealing, to reduce heat loss and gain.
- High-efficiency heating, ventilation, and air conditioning (HVAC) systems, including advanced controls and variable speed drives, to improve energy efficiency and occupant comfort.
- Efficient lighting systems, including LED fixtures and advanced lighting controls, to reduce electricity consumption and improve lighting quality.
- Renewable energy systems, such as solar photovoltaic panels or wind turbines, to generate electricity on-site and reduce reliance on grid-supplied energy.

4.6.2 Industry standard practice

In the context of code baselines, "industry standard practice" refers to the established and widely accepted methods for determining the minimum energy efficiency requirements for buildings based on building codes, standards, and guidelines. These methods are typically developed and maintained by industry associations, professional organizations, and government agencies, and they are based on extensive research, analysis, and consensus-building processes.

For example, in the United States, the International Energy Conservation Code (IECC) is a model building code that sets minimum energy efficiency standards for residential and commercial buildings. The IECC is developed by the International Code Council (ICC), a nonprofit organization that brings together building industry professionals, government officials, and other stakeholders to develop consensus-based codes and standards.

Similarly, ASHRAE publishes a set of standards and guidelines for energy efficiency in buildings, including Standard 90.1, which sets minimum requirements for energy efficiency in commercial buildings.

In utility energy efficiency programs, "industry standard practice" for code baselines typically involves using the building code or energy code that was in effect at the time of construction or renovation, as well as any applicable state or federal standards or guidelines. This approach ensures that the code baseline reflects the minimum energy efficiency requirements that were in place at the time of implementation.

4.6.3 Industry standard practice in energy-savings performance contracting

The decision to consider or not consider industry standard practice when calculating the impact of energy management activities in energy-savings performance contracting (ESPC) can depend on various factors, including the ESCO and its client's goals and priorities, the regulatory environment, and any contractual requirements.

From the perspective of the ESCO, using industry standard practice can help ensure that impact calculations are credible and defensible, which can help win contracts and maintain a positive reputation in the industry. It can also keep calculations consistent and transparent, simplifying the processes of comparing different projects' results and identifying best practices. From the client's perspective, using industry standard practice can provide reassurance that quantifications of the impacts of energy management activities are accurate and that the expected benefits will be realized (Figure 4.3). However, there may be situations where industry standard practice is less appropriate or beneficial, such as when the project involves innovative or customized measures not well represented by industry standards. In such cases, the client and ESCO may need to work together to develop alternative methods for calculating impacts that are appropriate for the specific project.

4.6.4 Expected useful life (lifecycle costing)

Expected useful life is a critical factor in lifecycle costing because it determines the duration over which the project costs and benefits will be realized.

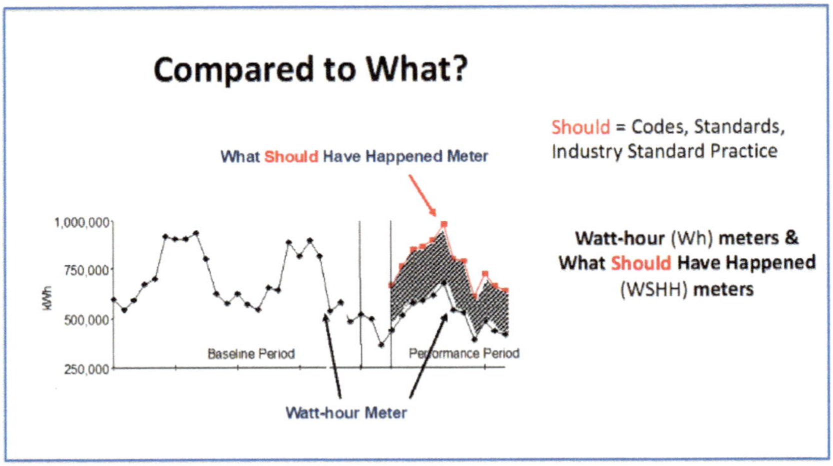

Figure 4.3 Industry standard practice as counterfactual.

In lifecycle costing, the total cost and value of a project is calculated over its expected useful life, which includes not only the initial installation costs but also the ongoing costs of operation, maintenance, and replacement.

If the expected useful life of a measure is short, it may not be cost-effective to implement the measure, as the impact generated over the short term may not offset the initial installation costs. On the other hand, if the expected useful life is long, the measure may be more cost-effective, as the impact generated over the life of the measure may offset the initial installation costs.

For example, suppose an LED lighting system is expected to have a useful life of 10 years and costs $10,000 to install, but will reduce costs related to energy consumption by an estimated $2000 per year. Over the 10-year life of the system, the estimated total cost reduction will be $20,000, offsetting the initial installation costs.

By considering expected useful life in lifecycle costing, the ESCO and the client can make more informed decisions about which measures to implement and how to allocate resources. They can also evaluate different financing options, such as whether to finance the project with a lump sum payment or recurring payments over time. Overall, expected useful life is a critical factor in lifecycle costing and is essential for ensuring that an ESPC achieves long-term financial and energy impact goals.

4.6.5 Dual baselines

The issue of expected useful life and code baselines can lead to the concept of dual baselines in utility energy efficiency programs. Dual baselines occur when there are different baseline assumptions for a measure over the reporting period.

For example, a utility energy efficiency program might include measures that are subject to upcoming code requirements (LED lights may be required in the near future). In this case, there may be one baseline for the current code and another baseline for the newer, more stringent code in the near future.

While rare, dual baselines are yet another concept that the M&V professional may encounter and needs to be able to explain.

4.6.6 First in, first out (FIFO)

FIFO (first in, first out) is a method used to assess the relative contributions of each ECM in a multiple-ECM project. In this method, the energy impacts associated with each ECM are calculated sequentially, in the order they were implemented. The method consists of the following steps:

1. List all the ECMs in the order in which they were implemented.
2. Calculate the energy impact of the first ECM implemented.
3. Calculate the energy impact of the second ECM implemented, with the first ECM already in place.
4. Continue this process for each subsequent ECM, adding the energy impact associated with each new ECM to the cumulative energy impact of the previous ECMs.
5. Evaluate the relative contributions of each ECM by comparing the incremental energy impact associated with each ECM to the cumulative energy impact of all the previous ECMs.

Using the FIFO method can help identify which ECMs are contributing the most to the overall energy impact of a project. But remember that the energy impact associated with each ECM can be affected by many factors, including the interactions between the ECMs, changes in operating conditions, and the variability of energy use over time. Therefore, you should employ additional methods, such as regression analysis or engineering estimates, to further refine the energy impact estimates and evaluate each ECM's relative contribution.

4.6.7 Order of operations and impact on individual ECM accounting

The issue of order of operations refers to the sequence in which energy conservation measures (ECMs) are implemented and how this can impact the quantification of energy management activity impacts. In many cases, the order of implementation can have a significant effect on the overall impact achieved, as well as the attribution of those impacts to individual ECMs.

For example, consider a building retrofit project that involves the installation of new lighting fixtures and an HVAC system upgrade. If the lighting fixtures are installed first, the baseline energy consumption for the building will be reduced, which may result in a lower energy consumption baseline for the HVAC system. As a result, the energy impact associated with the HVAC system upgrade may be under-estimated if the baseline assumptions do not properly account for the new lighting fixtures.

To mitigate the effect of order of operations on ECM accounting, you should consider the potential interactions between different measures. This may involve conducting a comprehensive energy audit to identify all of the energy-saving opportunities and determine the optimal sequence of ECM implementation.

4.6.8 Attribution, net to gross, and additionality

In a utility program, the net-to-gross (NTG) ratio indicates what portion of energy impacts can be directly attributed to the program itself. In essence, it is a measure of causality. It is calculated by dividing the net impact (the change in energy consumption attributable to the program) by the gross impact (the total observed change in energy consumption, regardless of cause). The resulting ratio ranges from 0 to 1, where 1 means that the program is directly responsible for all observed impacts, and 0 means that it is directly responsible for none.

The NTG ratio helps utilities measure the true effectiveness of their energy efficiency programs, revealing how much of the change in energy consumption is actually due to their intervention. This allows them to refine their programs, provide accurate information to stakeholders, and ensure that their efforts (and money) are being well spent.

4.6.9 NTG in ESPC

NTG is not typically used in the ESPC context. In an ESPC, the contractor's payment is often based on "guaranteed savings," typically calculated based on a baseline period before the implementation of the energy efficiency measures. The contractor's fee is then adjusted based on the actual impact of the energy management activities compared to the baseline period.

4.7 Using the Baseline Model and Information in the Reporting Period

This is what makes M&V, M&V.

This is the implementation of the "What Would Have Happened Meter."

This is where the magic happens – and where malfeasance can play a role.

If you have done everything correctly up until now, this should be "easy." But if you have not, this is where disagreements and disputes can arise.

Remember, it is always possible to take measurements during the reporting period. But it is impossible to know precisely what would have occurred had we not implemented energy management. For that, we need to have used baseline data to build a model. We need to have obtained agreement from stakeholders on the type of the model and the process by which it would be "updated" during the reporting period. We need to have a complete inventory of baseline data, should the need arise to make a non-routine adjustment. All of that should have been documented in an M&V plan. And if that plan is well constructed, the process of updating the baseline model will proceed without issues.

4.7.1 Terminology time: adjustments

As we have seen, the general concept of implementing the counterfactual method is not unique to M&V. And it is perfectly standard for complex domains like M&V to create terms that allow for the communication of complex ideas. Such is the case with the word "adjustments" in M&V.

Other words could replace "adjusting" without changing the meaning: "updating," "running," "altering," "revising," and "tweaking," for example. But let us first clear up one language issue. Baselines are *not* adjusted. The baseline is all the data collected to document the baseline conditions. Once documented, the baseline is fixed, in the sense of "set." However, the baseline model derived from the baseline can be adjusted, tweaked, etc.

M&V documents describe two types of adjustments. The terms "routine" and "non-routine" adjustments refer to the adjustments made to baseline energy models to account for changes in conditions such as weather, occupancy, and production levels. "Routine" adjustments are made for predictable changes in these conditions that were identified as part of building the baseline model.

In contrast, "non-routine" adjustments are made for unexpected changes, such as equipment failures or unplanned shutdowns. For these, you will need to have the static factors you collected as part of the baseline.

The counterfactual method, which compares actual energy consumption to a counterfactual scenario where the facility or system is assumed to have operated under baseline conditions, is standard in M&V and other disciplines such as economics and statistics. In these disciplines, similar terminology may be used to describe adjustments made to account for changes in conditions.

For example, in economics, the term "ceteris paribus" (meaning "all else being equal") is often used to describe the assumption that all variables except the one in question are held constant. In this context, adjustments made for changes in other variables could be described as "routine" or "non-routine" depending on their predictability.

In statistics, the term "covariate adjustment" describes the process of adjusting for confounding variables that may affect the relationship between two variables being studied. In this context, adjustments made for predictable confounding variables could be described as "routine," while adjustments made for unexpected confounding variables could be described as "non-routine."

While the specific terminology may differ across disciplines, adjusting for changes in conditions using a counterfactual method is a common approach to evaluating the effects of a particular variable of interest. Terms such as "routine" and "non-routine" adjustments can help clarify the predictability of these changes and aid in interpreting results.

4.7.2 Routine adjustment – known relationship (coefficient, parameter)

The M&V plan will include the details of what data will be collected and used to tweak the baseline model in the reporting period. This step in the overall process is expected and hence "routine." The "adjustment" typically involves executing a simple pre-determined calculation.

4.7.3 Non-routine: unknown and without previous calculation

Any event of sufficient magnitude (within the measurement boundary) not anticipated in the M&V plan and codified as a routine adjustment is a non-routine event. This can quickly become a concern with broad measurement boundaries and complex facilities.

There are two primary ways of identifying non-routine events: observing and analyzing energy consumption data or surveying the facility.

Non-routine Event Detection through Data

Non-routine events in M&V can be identified by analyzing the energy consumption data and looking for deviations from the expected or predicted energy consumption. Various non-routine events, such as equipment failures, production disruptions, or changes in occupancy patterns, could cause these deviations.

To identify non-routine events using energy consumption data, M&V practitioners typically use statistical techniques to model the expected or predicted energy consumption during a given period.

For example, if the predicted energy consumption during a particular period is 1000 kWh (after adjusting for routine events) and the actual energy consumption is only 800 kWh, this could indicate a non-routine event such as equipment failure or a production shutdown. On the other hand, if the predicted energy consumption is 1000 kWh and the actual energy consumption is 1100 kWh, this could indicate a non-routine event such as an unexpected increase in occupancy or production.

Once potential non-routine events are identified, M&V practitioners would need to investigate the cause of the deviation and determine whether an adjustment to the energy consumption data is necessary.

The non-routine adjustment must consider the event's cause and the stakeholders' responsibility. For example, if the deviation was caused by an equipment failure related to the ECM, the implementor may need to make good on a guarantee. If the deviation resulted from a change of operations attributable to the facility, the implementor should not be penalized. If the event was caused by forces outside the control of either party, such as extreme weather or other natural causes, the contract or program rules should have a clause to cover such a possibility.

Non-routine event detection in M&V is similar to fault detection and anomaly detection in other fields. All three involve the identification of deviations or anomalies from expected or normal behavior. In fault detection, the goal is to identify when a system or equipment is not functioning as intended.

This is typically done by monitoring system variables such as temperature, pressure, or flow rate and looking for deviations from expected or normal values. Detecting these deviations can help identify potential equipment failures or malfunctions and trigger maintenance or repair actions.

Similarly, anomaly detection aims to determine when a system or process is not behaving as expected. This is typically done by monitoring data from sensors or other sources and looking for patterns or deviations from expected or normal behavior. Anomaly detection can be used in various fields, such as cybersecurity, finance, and healthcare, to identify unusual or unexpected behavior that may indicate fraud, cyber-attacks, or medical anomalies.

In all three cases, detecting these deviations or anomalies can help identify potential problems or opportunities for improvement. By identifying and addressing these issues, it is possible to improve the performance, efficiency, and reliability of systems and processes.

However, a highly sensitive detection mechanism can quickly become problematic. The sensitivity of the fault detection method is typically set using a threshold value. The threshold value represents the maximum allowable deviation from the expected or normal values, above which the fault detection method will flag the data as a non-routine event. The threshold value can be set based on several factors, including:

- **Historical data:** One approach is to use historical data to determine the typical range of energy consumption or other relevant variables for the system. The threshold value can then be set based on the historical range of values, such as a certain number of standard deviations from the mean.
- **Engineering calculations:** Another approach is to use engineering calculations to determine the expected or normal values for the system. The threshold value can then be set based on the expected deviation from the normal values.
- **Stakeholder input:** Input from stakeholders, such as building occupants or production managers, can also be used to set the threshold value. These stakeholders may have insights into typical energy consumption patterns or production processes that can be used to inform the threshold value.

Once the threshold value is set, the sensitivity of the fault detection method can be adjusted by increasing or decreasing the threshold value. A lower threshold value will increase the method's sensitivity, which means that more non-routine events will be detected but may also result in more false alarms. Conversely, a higher threshold value will decrease the method's sensitivity, meaning fewer non-routine events will be detected, but may also result in missed events.

The sensitivity of the fault detection method should be evaluated periodically to ensure that it detects non-routine events accurately while minimizing false alarms. The threshold value may need to be adjusted based on changes in the system, such as changes in occupancy patterns or production processes.

EVO produced an application guide on non-routine adjustments.[25]

Non-routine Event Detection through Physical Inspection

While patterns in energy consumption may reveal non-routine events, that type of analysis cannot usually be used to determine the exact nature of the event. For that, it is typically necessary to survey the facility. For example, a non-routine event such as a duct leak may not be immediately apparent from the energy consumption data. A physical inspection of the ducts may be necessary to identify the cause of the leak and take corrective action.

Installing monitoring devices is another way to detect non-routine events physically. Non-routine events such as equipment failures or malfunctions can cause changes in the physical characteristics of the system or equipment that these sensors or meters can recognize. For example, if a non-routine event such as a bearing failure occurs in a motor, it can cause changes in the vibration characteristics of the motor. By monitoring the vibration levels of the motor using vibration sensors, it is possible to detect the presence of the non-routine event and take corrective action.

Responsibility for Documentation

The responsibility for documenting non-routine events typically falls on the M&V practitioner or the energy manager responsible for monitoring and analyzing energy consumption data. Whoever is responsible should maintain a log or record of non-routine events, including the event's date and time, the nature of the event, and any corrective actions taken. This information can help identify the cause of the non-routine event and determine whether any adjustments to the energy consumption data are necessary.

4.7.4 Non-routine adjustments

Once the event has been identified, the M&V practitioner or energy manager must estimate the event's impact on energy consumption – the non-routine

[25]https://evo-world.org/en/news-media/evo-news/1195-release-of-the-ipmvp-application-guide-on-non-routine-events-and-adjustments

adjustment. This can be done using various methods, such as energy modeling, engineering calculations, or empirical data.

Anticipated NRAs

In some cases, non-routine events can be anticipated and planned for, which means that the impact of these events on energy consumption can be estimated and factored into the baseline energy model. This can be done by adjusting the baseline energy model to reflect the expected impact of the event before the event occurs.

For example, if there is an expectation of future facility operational changes (reduced or increased hours of use), the M&V plan may include a method for adjusting the baseline model when such events occur. The analysis could be based on engineering calculations, energy modeling, or historical data from similar events. Agreeing on this analysis ahead of time can help eliminate disputes. However, it is important to note that not all non-routine events can be anticipated or planned for, and in these cases, adjustments will need to be made after the event occurs.

Ad-Hoc Non-routine Events (Not Anticipated)

For non-routine events that were *not* anticipated, several methods can be used to estimate the impact of the event on energy consumption and make non-routine adjustments:

- **Empirical data:** One method is to use empirical data from similar non-routine events that occurred in the past to estimate the impact of the current event. For example, suppose a production facility experienced a similar production disruption in the past. In that case, the M&V practitioner or energy manager can use historical energy consumption data from that event to estimate the impact on energy consumption for the current event.
- **Statistical analysis:** Another method is to use statistical analysis to identify patterns in the energy consumption data that may be related to the non-routine event. This can involve analyzing energy consumption data before and after the event to identify changes in energy consumption patterns and using statistical methods to estimate the event's impact on energy consumption.
- **Energy modeling:** Energy modeling can be used to simulate the impact of the non-routine event on energy consumption. This involves developing a model of the energy system and simulating the event's impact

on energy consumption using the model. This method can be helpful for complex systems where empirical data is limited.

- **Engineering calculations:** Engineering calculations can be used to estimate the impact of the non-routine event on energy consumption. This involves using engineering principles and calculations to evaluate the event's impact on energy consumption based on the system's characteristics and the event.

In general, the method used for estimating the impact of a non-routine event on energy consumption will depend on the nature of the event, the availability of data and resources, and the system's complexity. The M&V practitioner will apply judgment to select the most appropriate method based on the specific circumstances of the event.

NREs in the Baseline Period

Non-routine events can occur during the baseline period. In Section 4.5, "Fair and Relevant," we touched on the topic of ensuring that the baseline period covers a span of operations that are expected to be typical of how the facility is operated in the future, the reporting period. However, if there are events within the selected baseline period that are not expected to recur, they can be analyzed, and the energy consumption "removed' from the baseline model.

For example, if a building experiences a power outage during the baseline period, the energy consumption data during the outage period may be missing or significantly different from normal operating conditions. This can affect the accuracy of the baseline energy model and hence the calculation of the impacts of energy management activities during the post-implementation period.

To account for non-routine events during the baseline period, one could document and characterize any non-routine events and adjust the baseline energy consumption level accordingly, such as by removing data during the non-routine event period. Judgment is required.

4.8 Granularity and Load Shapes

The fundamental data in M&V is the time-series pattern of the baseline, adjusted throughout the reporting period. When conducting M&V, professionals must decide on the unit of time they will use for the time-series data. The specific application, metering available, and the level of accuracy needed should inform this choice. Monthly utility bill data is still commonly used in many contracts. But with the increase in data available from "smart" meters, there is a growing interest in the use of more granular data. The challenge for an M&V professional is to balance the costs of collecting and analyzing data against the benefits of enhanced understanding of the system.

The optimal frequency for data analysis purposes may be different from the frequency of the underlying data. Depending on the application, highly granular data can be aggregated up to a frequency that proves best for analysis purposes (low CV(RMSE)).[26]

In some cases, high frequency data is required to capture the effects of the ECM. For example, if the goal is to assess the impact of an hourly demand reduction program, hourly data may be sufficient. However, if the goal is to assess the impact of a more granular program, such as a residential smart thermostat program, 15-minute or even 5-minute data may be required to accurately capture the energy usage patterns.

[Sidebar]

AMI (advanced metering infrastructure) and M&V 2.0 are closely related, as AMI can provide the necessary data for M&V 2.0. M&V 2.0 is an advanced approach to M&V that leverages high-resolution data from advanced meters, smart building systems, and other Internet of Things (IoT) devices to provide more accurate and detailed analysis of energy consumption.

AMI allows for the collection of high-resolution energy usage data at much finer time intervals (e.g., every 15 minutes) than traditional monthly utility billing data. This high-resolution data can provide a more accurate view of energy usage.

M&V 2.0 is still evolving, and there is ongoing research and development to better integrate AMI and other IoT technologies into the M&V process. However, the availability of high-resolution energy usage data from AMI and other IoT devices will play an increasingly important role in the future of M&V.

4.8.1 Granularity leads to understanding load shapes[27]

Granularity and load shapes are both related to the level of detail in energy use data.[28]

---

[26]https://towardsdatascience.com/12-things-you-should-know-about-time-series-975a185f4eb2

[27]https://www.youtube.com/watch?v=7BHQfk6kvso

[28]https://www.nrel.gov/buildings/end-use-load-profiles.html

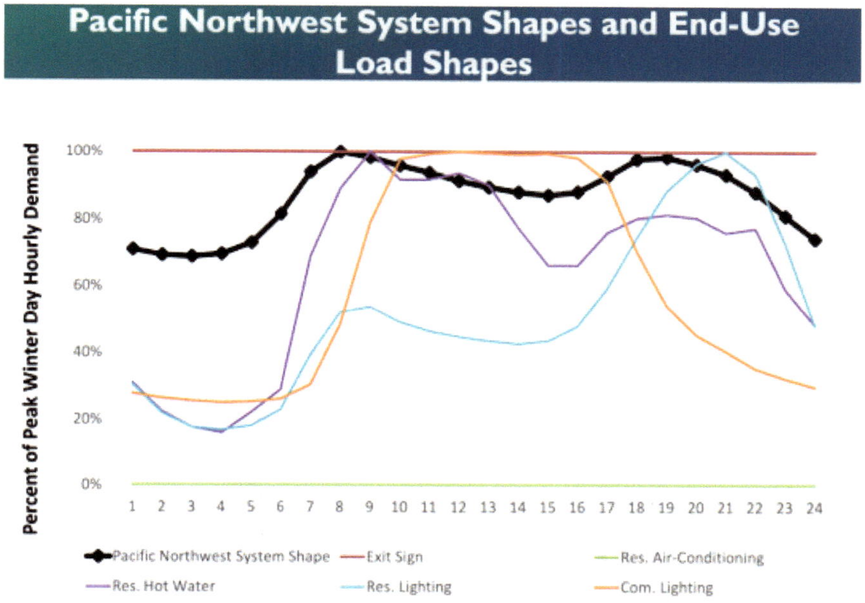

Figure 4.4 System hourly load shapes by end-use.

Granularity refers to the level of detail in the energy use data, which can vary from monthly utility bills to sub-hourly interval data. Granularity can impact the accuracy and precision of energy impact calculations, as more granular data can provide more detailed information about energy use patterns and the effectiveness of energy efficiency measures.

Load shape, on the other hand, refers to the pattern of energy use over time, typically daily or hourly. Load shape can impact the selection and implementation of energy efficiency measures, as measures that target specific times of day or specific equipment usage patterns can disproportionately reduce energy consumption.

Higher granularity data can provide more detailed load shape information, allowing for more precise and accurate calculations. However, collecting and analyzing this data can be more expensive and time-consuming and may require specialized software and expertise.

As always, the appropriate level of granularity and load shape detail will depend on the specific goals of the energy efficiency program and the available resources for data collection and analysis.[29]

[29]https://emp.lbl.gov/publications/time-varying-value-electric-energy

4.8.2 Getting a correlation: granularity matters

It is generally true that lower granularity data (e.g., daily, weekly, or monthly) can be more easily correlated to external variables such as weather or occupancy patterns, as the impact of these variables tends to be more visible at a higher level of aggregation.

For example, daily energy use data can be used to analyze the impact of weather patterns on energy use, as daily temperature fluctuations are more apparent than hourly temperature fluctuations. Similarly, weekly or monthly data can be used to identify longer-term trends in energy use, such as seasonal variations or changes in occupancy patterns.

However, while lower granularity data can help identify broad patterns and trends, it may not provide enough detail to accurately model the impact of specific energy efficiency measures or identify areas for improvement. In such cases, higher granularity data may be necessary to provide the level of detail required for accurate analysis and modeling.

In some situations, monthly data will (have to) suffice. In others, daily is preferable. And there are times when hourly data is appropriate. Selecting the most fitting time interval requires judgment – though, in some situations, a particular level of granularity will be required.

4.8.3 Assigning value to "savings"

The savings load shape can be used to determine the value of the impacts of energy management activities by comparing the actual energy consumption after the energy efficiency project with the energy consumption that would have occurred in the absence of the project, taking into account changes in energy use expected to result from the project.

The savings load shape allows one to disaggregate the overall energy "savings" into hourly or monthly increments. This allows for a more precise determination of the value of the impact of energy management activities, as different hours or months may have different electricity rates, time-of-use pricing, or demand charges.

To determine the value of "savings" using the savings load shape, the energy consumption data is multiplied by the corresponding energy rates or tariffs for each hour or month and then compared to the expected energy consumption and cost in the absence of the energy efficiency project.

The savings load shape is becoming increasingly important as underlying energy tariffs become more granular.[30,31]

4.8.4 Demand savings

Demand reduction is typically calculated using interval data that captures energy usage at regular intervals (e.g., every 15 minutes) over a specific period. This interval data is then used to calculate the average demand during that period, expressed in kilowatts (kW).

Monthly utility bills do not provide enough granularity to accurately calculate demand reduction because they only offer a single data point for energy usage over an entire month. However, estimating demand reduction using monthly data using a simplified approach that assumes a constant load shape throughout the day is possible. This approach, known as the peak demand method, involves identifying the maximum demand during the billing period and comparing it to a baseline period to determine the demand reduction. While this method is less accurate than interval data, it can provide a rough estimate of demand reduction and may be sufficient for some projects.

4.9 Impact Terminology

Until now in this book, we have assumed that the M&V process moves along, and the results are reported based on the actual conditions of the reporting period. This is indeed how most M&V projects proceed. The results from these projects are termed "avoided energy" and "avoided/cost." There are two other common methods for calculating impacts. "Normalized energy/cost" differs from "avoided" by generalizing the impact calculation to conditions other than what occurred in the reporting period. The other, "backcasting," is the practice of establishing a model of the reporting period energy and using it to calculate what the facility would have used under baseline conditions.

The use of the term "avoided energy/cost savings" or "normalized savings" is instantiated in the M&V guidance documents. As elsewhere in this book, we avoid using "savings." But in some cases, it is the term or art that an M&V professional must use to be understood.

---

[30]https://eta-publications.lbl.gov/sites/default/files/lbnl_webinar_timevaryingee_071017.pdf

[31]https://www.irena.org/-/media/Files/IRENA/Agency/Publication/2019/Feb/IRENA_Increasing_time_granularity_2019.pdf

And the use of the term "normal" in this situation also causes us to remember that many words in M&V can cause confusion. "Normal" is used in many different ways and contexts, each with a unique meaning. Be careful!

Let us investigate all three in a bit more detail.

4.9.1 Avoided energy/cost savings

This refers to the estimated impact of the energy efficiency project under the conditions experienced in the reporting period. The avoided energy/cost is calculated by comparing the actual energy consumption or cost after the project is implemented to the counterfactual (estimated energy consumption or cost that would have occurred in the absence of the project). This method accounts for changes in external factors that may affect energy consumption or cost, such as weather patterns or occupancy levels.

While this method is "accurate" by quantifying what occurs during the reporting period, it can under- or over-estimate long-term impacts if the reporting period conditions are abnormal. A particularly cool summer might reduce a chiller project's "savings" and under-predict its long-term impacts.

4.9.2 Normalized savings: energy/cost

Normalized savings allow parties to consider the impact of an energy project under a different set of conditions than in the reporting period. Normalized savings are calculated by creating baselines and reporting period energy consumption models. The routine adjustment for both models is an agreed-upon reference dataset. It could be long-term average weather (or pertinent independent variable) or any data that covers a scenario of interest.

4.9.3 Backcasting

Backcasting is a method used to estimate the impacts that would have been achieved if an energy efficiency project had been implemented under baseline conditions. This involves using the reporting period data to create a project model after implementing measures. That model can calculate what the retrofitted building would have used under baseline conditions – assuming baseline data is available. The resulting energy usage is compared to what was used – typically at an annual level – in the baseline period.[32]

[32]https://semhub.com/assets/documents/uploads/NWSEMC_MV_WG_Energy_Modeling_Method_Selection_Guide_2019.pdf

4.10 Contracting for Savings

When implementing energy management with ESPC, you should be aware of the general structure of the contract, as this impacts the role and importance of M&V.

Generally speaking, performance contracts take one of two forms: guaranteed savings or shared savings.[33] The fundamental difference between guaranteed savings and shared savings is in the way that financial risks are allocated between the parties involved.

Under a guaranteed savings model, the ESCO guarantees a certain amount of energy "savings" from the project and is responsible for any shortfall. The customer pays the ESCO a fixed fee for the project, based on the guaranteed savings, regardless of the actual impact of the intervention.

Under a shared savings model, the customer and the ESCO share the risks and benefits of the energy efficiency project. The ESCO is paid a percentage of the energy "savings" achieved, typically over a fixed period. If the impacts of the energy management activities are less than expected, the ESCO receives a lower payment, and the customer retains a larger share. Conversely, if the actual impacts are greater than anticipated, the ESCO receives a higher payment.

The choice between guaranteed and shared savings depends on several factors, including the project scope, the level of financial risk the customer is willing to accept, and the ESCO's capabilities and preferences.

Guaranteed savings models may be more suitable for projects with well-established, predictable forms of intervention. In contrast, shared savings models may be more appropriate for projects with less certain outcomes.

Shirly Hansen provided a chapter documenting the evolution of performance contracting.[34]

4.10.1 What does this mean for M&V?

The "guarantee" is only as good as the M&V.

Even in a guaranteed savings contract, there still needs to be good M&V to ensure the guarantee is being met and provide a basis for ongoing performance monitoring and energy system optimization.

The guarantee is typically set below the expected "savings." This provides a buffer to acknowledge uncertainty and reflects potential risks. So long as

[33]https://www.iea.org/reports/energy-service-companies-escos-2/esco-contracts

[34]Energy Project Financing – River Publishers

the results are "reasonable" – i.e., higher than the guaranteed amount – the project is a success. However, closer scrutiny may be required if the "savings" reported through M&V get closer to the guaranteed amount.

Meanwhile, in a shared savings contract, the savings for each period are split between the owner and the contractor, and each calculation receives even greater scrutiny. This is why having a robust M&V plan in place and good communication throughout the process is important.

4.11 Standard Weights and Measures

An M&V professional is often called upon to exercise judgment to create the optimal M&V plan for a particular project. But in some situations, the best solution is a standardized method or algorithm that requires adherence rather than creativity. For example, in a utility program with multiple vendors implementing projects and performing M&V, it might be better to agree on a common methodology for all the projects to allow for comparisons across the portfolio.

In addition to consistency, standardized methods and algorithms offer several other benefits:

- **Transparency:** They are publicly available, which increases the transparency of the M&V process, helps build trust among stakeholders, and improves the credibility of the results.
- **Cost-effectiveness:** They are often developed and maintained by industry associations or government agencies and are freely available to the public – for example, CalTrack, OpenEE Meter, and the DOE energy modeling tools DOE2 and EnergyPlus. They also tend to be designed to be used without extensive customization.
- **Transferability:** They can easily be transferred between projects and organizations, making replicating successful projects and scaling up energy efficiency programs easier.

4.12 Measurement Site vs. Source Issue

While M&V is usually focused on the impacts of energy management at an end-use facility, when a distribution grid serves that facility, there are also impacts upstream. The U.S. Environmental Protection Agency (EPA), ISO 50001, and the U.S. Energy Information Agency (EIA) have all considered how to account for these impacts.

4.12.1 EPA

In carbon emissions accounting, "scope" refers to the boundaries or limits for measuring and reporting greenhouse gas (GHG) emissions. There are three scopes:

- *Scope 1 emissions* are direct emissions from sources owned or controlled by the reporting organization. Examples include emissions from on-site boilers, vehicles, and equipment.
- *Scope 2 emissions* are indirect emissions from the generation of electricity, heat, or steam purchased by the organization. These emissions are considered indirect because they are produced by utility companies rather than the reporting organization.
- *Scope 3 emissions* are all other indirect emissions that occur in the reporting organization's value chain, such as emissions from the production of raw materials, the transportation of goods, and the disposal of waste.

Organizations typically focus on scopes 1 and 2 as they have more control over these emissions and can take more direct actions to reduce them. Scope 3 emissions are generally more challenging to address as they are outside an organization's direct control. Still, they can significantly impact a company's overall carbon footprint, and organizations are encouraged to consider these emissions in their sustainability reporting.

4.12.2 ISO 50001

ISO 50001, the international standard for energy management systems, uses several terms to describe its requirements and processes, including the following:

- **Energy review:** An evaluation of the organization's energy use, including an analysis of energy consumption, costs, and opportunities for improvement.
- **Energy performance indicators (EnPIs):** Metrics used to monitor and evaluate the organization's energy performance. EnPIs are established during the energy review process and may include metrics such as energy consumption per unit of production, energy intensity, and specific energy consumption.
- **Energy management objectives and targets:** Goals established by the organization for improving its energy performance. These objectives and targets are based on the results of the energy review and may include

goals for reducing energy consumption, improving energy efficiency, or reducing greenhouse gas emissions.

- **Energy baseline:** The starting point from which an organization's energy performance is measured. The energy baseline is established during the energy review and provides a reference point for tracking progress toward energy management objectives and targets.
- **Energy management action plan:** A plan that outlines the steps that the organization will take to achieve its energy management objectives and targets. The action plan may include implementing energy-efficient technologies, process improvements, and changes to energy management practices.
- **Energy management system (EnMS):** An organization's systematic approach to managing its energy use and reducing its energy consumption.
- **Continuous improvement:** The ongoing process of refining and improving the organization's energy management practices and procedures. This includes regularly monitoring energy performance, identifying opportunities for advancement, and implementing changes to achieve energy management objectives and targets.

4.12.3 EIA

The EIA – the statistical arm of the U.S. Department of Energy, responsible for collecting and analyzing data on energy production, consumption, and prices – distinguishes between "site" and "source" energy. "Site" energy refers to the energy consumed at a specific location, such as a factory, office building, or household. This is similar to the EPA's scope 1 emissions, which include direct emissions from sources owned or controlled by an organization.

By contrast, "source" energy is the total energy used to generate, transmit, and distribute energy to a site. This is similar to the EPA's scopes 2 and 3 emissions, which include indirect emissions from the generation of electricity, heat, or steam purchased by an organization and all other indirect emissions that occur in a company's value chain.

Other EIA terminology includes:

- *Energy conversion factors* are used to convert between different units of energy, such as from Btu to kWh. The EIA provides energy conversion factors for various energy sources and uses.

- *Energy intensity* is a measure of energy efficiency calculated as the energy consumed per unit of economic activity, such as per dollar of gross domestic product (GDP) or per unit of production. Energy intensity is expressed in physical units of energy per unit of economic activity.
- *Energy balances* track energy flow into and out of a system, such as a country or a region. Energy balances include data on energy production, consumption, imports, exports, and inventory changes.
- *Energy flow diagrams* visualize the energy flow in a system. Energy flow diagrams show the sources and types of energy consumed, as well as the energy transformation processes used to generate energy.
- *Energy efficiency* is the ratio of useful energy output to the energy input required to produce it. Energy efficiency is expressed as a percentage and is used to compare the energy performance of different systems or processes.

4.13 Cost

As mentioned earlier, M&V is the accounting system for energy management. For most operations, the accounting function is an overhead cost. But in any well-run organization, the information obtained through accounting can and will be used to make decisions that improve overall operations. In any case, M&V activities do incur a cost, and especially with small projects, that can be prohibitive. In other words, M&V activities need to be scaled to meet the needs and budget of each project.

In some cases, like a sampled project from a utility program portfolio, the M&V costs may exceed the impacts of that one project. But if the results are used to extrapolate across a large portfolio, it may be well worth the expense. In other words, the flip side of M&V costs is the benefit obtained from M&V activities – a classic cost–benefit analysis.

In this section, we will dive deeper into the types of decisions that you will face as an M&V professional when balancing the costs and benefits of a specific M&V plan.

4.13.1 Categories of costs to consider

Here is a list of potential costs you should keep in mind when developing an M&V plan. Amounts will vary depending on the specific project, scope, and location.

- Personnel – e.g., project managers, data analysts, and field technicians

- Equipment costs – for purchasing or renting equipment to measure energy use, such as data loggers, meters, and sensors
- Software costs – for purchasing or licensing software to process and analyze data, such as energy management systems and data analysis tools
- Travel and lodging costs
- Administrative costs – for project management, documentation, and reporting
- Contingency costs – for unforeseen events or issues that may arise during implementation, such as equipment failure or delays in data collection
- Legal costs – for legal advice or assistance in drafting the M&V plan, reviewing contracts, or resolving disputes

Two good resources that are available to assist with M&V cost planning are the U.S. Army M&V Costing Toolkit[35] and the NSW Office of Environment and Heritage – M&V Excel planning tool.[36,37]

Besides the obvious value in quantifying the impacts of energy management activities for contractual or programmatic purposes, M&V can also lead to improved operational performance through observation and feedback, detection and correction of faults, risk mitigation through observation, increased stakeholder confidence, and reduced uncertainty and increased confidence. Granted, these outcomes are not always realized in every project, nor are they always needed – but it is difficult to say for which projects they will be valuable. In this sense, M&V can be thought of as a kind of insurance. Having information available to resolve disputes may be much less costly than litigation.

4.13.2 Sharing costs with others

Energy and baseline data are critical to many aspects of an energy management activity – not just M&V. Ideally, the same data would never be collected more than once, but in practice, it happens often. As an M&V professional, you should be alert to any opportunities to use data from other participants in the energy management process and find "customers" for the data collected

[35] https://oaktrust.library.tamu.edu/handle/1969.1/2050

[36] https://www.environment.nsw.gov.au/resources/energyefficiencyindustry/120990bestpractice.pdf

[37] https://apps.environment.nsw.gov.au/resources/energyefficiencyindustry/120999planningtoolexample.xls

through M&V activities. This way, M&V costs can be shared with the audit, commissioning, test and balance, and other teams. By coordinating efforts and sharing costs, the overall project budget can be optimized while ensuring that all necessary tasks are completed effectively.

4.13.3 General guidance

Unless you have a specific budget or means of assessing the benefit of increased M&V, keep M&V costs as low as stakeholders will allow.

The cost of M&V can range from a few thousand dollars for a simple project to several hundred thousand dollars – or more – for a complex project demanding a high level of detail and rigor. As a rule of thumb, M&V costs typically range from 1% to 5% of the total project cost. However, this can vary widely depending on the project and approach.

4.13.4 Optimizing M&V expenditures

As we will see in the next section, all energy management activities involve a degree of uncertainty. Many financial domains have developed strategies for dealing with uncertainty by incorporating risk premiums. A risk premium is an additional return or compensation that investors or participants demand in exchange for taking on a higher level of risk in an investment or financial transaction.

To the degree that M&V can reduce uncertainty, one can justify M&V costs as a necessary investment in risk reduction and uncertainty management. A risk assessment can help to identify potential sources of error or bias in the M&V process, such as incomplete or inaccurate data, faulty assumptions, or measurement errors. An uncertainty reduction algorithm can be used to predict the effect of various sources of uncertainty on the estimated impact of energy management activities and to identify the most effective strategies for reducing uncertainty. For example, the algorithm might identify the need for additional data collection or more frequent measurements to minimize uncertainty or suggest changes to the M&V plan to improve accuracy.

Using Monte Carlo

Monte Carlo simulation is a valuable tool for identifying potential sources of error or uncertainty in M&V analysis. It involves generating a large number of random samples of input data based on probability distributions and using these to estimate the distribution of the output. By varying the input parameters, the model can be used to identify which parameters have the greatest impact on the output and, therefore, which parameters need to be monitored most carefully – after considering the costs that this monitoring would entail.

Monte Carlo simulation can also help identify areas where additional data collection or analysis may be needed to reduce the uncertainty associated with the M&V analysis. By understanding which input parameters have the greatest impact on the output, resources can be focused on improving the accuracy and reliability of those parameters. For example, if weather data is found to be a significant source of uncertainty in the analysis, additional weather monitoring equipment may be installed, or more detailed weather data may be collected to reduce the uncertainty associated with the weather assumptions.

* * *

To develop an optimization tool that minimizes uncertainty or risk, one must consider factors such as the desired confidence level, the consequences of incorrect M&V, and the costs associated with reducing uncertainty. The tool should be designed to balance the costs of M&V against the potential risks and uncertainties associated with not performing adequate M&V.

The optimization algorithm would need to incorporate variables such as the costs of M&V activities, the probability of errors in impact calculations, and the potential consequences of those errors. It would also need to consider the relative importance of different M&V activities, such as the value of on-site inspections vs. data analysis.

The tool's output could be a recommended budget for M&V activities that provides the desired confidence level in the impact calculations while minimizing the overall risk and uncertainty. The tool could be updated as additional data is collected and the project progresses, allowing for ongoing optimization of M&V activities.

Example

Consider a facility whose ownership plans to implement several energy efficiency measures and wants to develop an M&V plan to assess the measures' effectiveness. There is a budget of $50,000 to spend on M&V activities.

To minimize the risk and uncertainty associated with the M&V plan, an optimization tool could be developed that considers the cost of various M&V activities and the potential reduction in uncertainty that each activity could provide.

The optimization tool might start by estimating the level of uncertainty associated with the impact estimates for each energy efficiency measure. For example, the uncertainty associated with a lighting retrofit might be estimated to be 20%, while the uncertainty associated with an HVAC upgrade might be estimated to be 15%.

Next, the optimization tool could calculate the cost of various M&V activities that could be used to reduce uncertainty. For example, installing submetering equipment to measure energy consumption in different areas of the facility might cost $5000, while conducting a detailed energy audit might cost $10,000.

The optimization tool could then use a Monte Carlo simulation to determine which combination of M&V activities would result in the greatest reduction in uncertainty while staying within the $50,000 budget. The simulation might involve running hundreds or thousands of scenarios, each with a different combination of M&V activities, and calculating the resulting uncertainty reduction and cost.

Based on the simulation results, the optimization tool could provide a recommended M&V plan that minimizes the overall uncertainty of the estimates while staying within the facility's budget.

For example, the tool might recommend installing submetering equipment and conducting a detailed energy audit for a total cost of $15,000, resulting in an estimated uncertainty reduction of 10% for the lighting retrofit and 12% for the HVAC upgrade.

Or another scenario –

In this one, the cost of M&V would not be fixed and would depend on the desired level of risk reduction. The value of M&V would be a function of the risk tolerance of the project participants, as they want to ensure a certain degree of impact at 80% confidence, one tail.

To develop an optimization tool in this scenario, the first step would be to identify the sources of uncertainty and risk associated with the project. This could include factors such as weather variability, occupancy patterns, equipment performance, and changes in building use.

Next, a probabilistic model could be developed to simulate the significance of these sources of uncertainty for the expected impacts of energy management activities. This model could generate a distribution of possible outcomes, which could be analyzed to determine the probability of meeting the target impact level at various levels of investment in M&V.

Using this information, an optimization algorithm could be developed to determine the optimal amount to spend on M&V to achieve the desired level of risk reduction. This would involve balancing the cost of M&V against the expected value of the risk reduction achieved.

For example, if the project participants were willing to accept a 20% chance of not achieving the target impact level, the optimization tool could be used to determine the minimum investment in M&V required to achieve an 80% confidence level for meeting the target.

The cost of M&V could then be compared to the expected value of the risk reduction achieved to determine if the investment in M&V is justified. If the cost of M&V is less than the expected value of the risk reduction achieved, the investment would be considered a good value. If the cost of M&V is greater than the expected value of the risk reduction achieved, the investment may not be justifiable.

Developing an optimization tool to minimize uncertainty or risk in M&V would involve input from several stakeholders. This would include the following:

- Energy efficiency program managers and staff would supply information on the specific project goals, performance metrics, and the level of risk that the program could tolerate.
- M&V professionals would advise on the "amount" or extent of M&V required to achieve the desired confidence level.
- Data analysts would give input on the type and granularity of data needed to reduce uncertainty.
- Financial analysts would offer insights on the value of the impacts of energy management activities, the cost of M&V, and the potential financial risks associated with the project.
- Software developers would design and develop the optimization tool to meet the project's requirements.

Required skillsets and areas of knowledge would include:

- M&V
- Statistical analysis and modeling
- Data analysis and visualization
- Financial analysis and risk management
- Software development and optimization

* * *

Actuarial tables[38] could be used to estimate the probability of various outcomes based on historical data or assumptions. For example, one could create actuarial tables that estimate the likelihood of achieving energy management impacts of a certain magnitude based on various factors, such as the specific ECMs being implemented, the size and type of the facility, and the weather conditions in the region. These tables could be based on historical data from similar projects or could be developed using simulation models.

Bayesian methods could be used to update the actuarial tables as new data becomes available. For example, if the actual impact achieved in a project is known, this information could be used to update the estimated probability of achieving similar impacts in future projects. The actuarial tables could also be used to estimate costs, which could help determine the optimal amount to spend on M&V or other project components.

4.14 Using Statistics to Communicate Uncertainty and Related Topics

Remembering that the M&V professional's main job is to ensure clear and consistent communication throughout the project, establishing a common understanding of the concept of uncertainty in M&V is critical to a successful settlement.

Besides "savings," uncertainty is among the most misunderstood but potentially valuable concepts.

This section reviews the main concepts and terminology used to communicate uncertainty. The reader can find details on these topics in any number of introductory or advanced texts. The critical issue for the M&V professional is ensuring that all stakeholders are aware of the reality of uncertainty and are comfortable with its management.

4.14.1 Statistics

Statistics provides a framework for analyzing and interpreting data, quantifying uncertainty, and evaluating the reliability and credibility of estimated energy management activity impacts.

One of the key roles of statistics in M&V is to help distinguish between random variation and real impacts on energy consumption patterns. Statistical techniques such as hypothesis testing and regression analysis allow us to

[38] https://www.sciencedirect.com/science/article/abs/pii/S0301421503003768

identify whether observed changes in energy consumption are statistically significant and can be attributed to the energy conservation measure in question or whether they are due to random variation ("noise") or other factors.

Statistics can also be used to quantify uncertainty. Tools and techniques such as confidence intervals, Monte Carlo simulations, and sensitivity analyses can help us assess the range of possible outcomes and the degree of confidence we should place in the estimate.

Similarly, we can evaluate the reliability and credibility of M&V results through statistical techniques such as goodness-of-fit tests, model validation, and diagnostic checks, which can help assess the accuracy and appropriateness of the statistical models and assumptions used in the analysis.

Descriptive Statistics

We cannot cover even a slice of the vast body of statistical techniques that might be of interest to M&V analysts. But there are a few simple statistical concepts that every M&V professional should master. The first is descriptive statistics.

> "Descriptive statistics serve as a fundamental step in the data analysis process, allowing analysts to gain insights into the nature of the data before more advanced statistical analyses are performed. They provide a foundation for making informed decisions and drawing conclusions based on the observed characteristics of the data."

The beauty of descriptive statistics is that any two people who apply descriptive statistics to a set of data should arrive at the same answer. This is a first step in building confidence.

Parameters used in Descriptive Statistics

Some of the parameters frequently employed in descriptive statistics include the following:

- **Mean:** The average of a set of values.
- **Median:** The middle value in a set of values when they are arranged in numerical order.
- **Mode:** The value that occurs most frequently in a set of values.
- **Range:** The difference between the maximum and minimum values in a set of values.

- **Interquartile range:** The difference between the third and first quartiles of a set of values and a measure of the spread of the middle 50% of the data.
- **Variance:** A measure of the spread of the values around the mean.
- **Standard deviation:** The square root of the variance and a commonly used measure of the spread of the values around the mean.
- **Skewness:** A measure of the degree of asymmetry in a set of values.
- **Kurtosis:** A measure of the degree of peakedness or flatness in a set of values.
- **Coefficient of variation:** A measure of the relative variability of a set of values, expressed as a percentage of the mean.

Example

Suppose you have a group of 10 motors and want to estimate the coefficient of variation (CV) of their power output. You measure the power output of each motor and record the following data (in kW):

5.6, 4.9, 6.2, 5.8, 5.5, 4.7, 5.9, 6.1, 5.2, 5.4.

To estimate the CV, you can use the following formula:

CV = (standard deviation $\div$ mean) $\times$ 100%.

First, you calculate the mean and standard deviation of the data:

Mean = (5.6 + 4.9 + 6.2 + 5.8 + 5.5 + 4.7 + 5.9 + 6.1 + 5.2 + 5.4) $\div$ 10 = 5.48 kW

Standard deviation = 0.459 kW.

With these values, you can then calculate the CV:

CV = (0.459 $\div$ 5.48) $\times$ 100% = 8.38%.

So, the estimated coefficient of variation for this group of motors is 8.38%.

Sampling Basics

Sampling is the second statistical concept that is often applied in M&V.

Sampling allows us to estimate population characteristics based on a subset of the population, without measuring every item or individual. In M&V, the "population" may consist of all the energy-consuming equipment or systems within a facility or a group of facilities. It may be impractical or infeasible to measure and analyze the energy consumption of every individual piece of equipment; so, instead, a representative sample can be selected and measured.

As long as the representative sample is chosen, this approach can reduce the time, cost, and resources required for M&V while still providing high accuracy and confidence in the results.

Unlike descriptive statistics, sampling can lead to a range of "answers."

In M&V, two samples can return different results, even if both of them were collected following standard procedures. This is because sampling is inherently probabilistic, and the results obtained from any sample are subject to random variation.

For example, suppose you are conducting M&V for a lighting retrofit project in a large office building. You decide to take two samples of the lighting fixtures, each with the same sample size. You follow standard procedures in selecting the samples and measuring the energy consumption before and after the retrofit.

If the two samples are truly random and representative of the population of lighting fixtures in the building, they should lead you to similar results. However, due to random variation, it is possible for the samples to return meaningfully different results. This could be due to factors such as variations in the age or condition of the fixtures, or measurement errors.

Importantly, if both samples followed standard procedures, the results obtained from each sample are equally valid and should be considered in the analysis. The difference in the estimates between the two samples indicates the uncertainty associated with the estimate.

Again, as long as the samples were selected and measured using standard procedures and appropriate statistical methods are used to account for uncertainty, it is acceptable if the two samples return different results. The key is to ensure that the M&V analysis is transparent and reasonably replicable so that others can understand the sources of uncertainty and make informed decisions accordingly.

4.14.2 Uncertainty

The main role of uncertainty in M&V is to quantify the level of confidence that can be placed in the M&V's results. Uncertainty is inherent in any measurement or estimation process. It reflects the fact that there are limitations to the accuracy and precision of the data and methods used. As an M&V professional, by quantifying uncertainty, you can communicate the level of confidence that can be placed in your estimates to help guide decision-making – both your own and that of other stakeholders.

- *Uncertainty* is the lack of knowledge or information about a particular outcome or event. It is the degree to which an outcome or event is unknown or unpredictable. Uncertainty is often measured using probability distributions, which assign probabilities to different possible outcomes. Probability distributions are often defined by using confidence intervals and confidence levels.

While performing M&V without uncertainty calculations is possible, it would not be a best practice. Awareness of uncertainty helps us recognize the limitations of our knowledge and understanding. It makes us more likely to ask critical questions, seek additional information and evidence, and engage in more robust analysis and decision-making processes. Ignoring uncertainty, on the other hand, can lead to a false sense of certainty and a lack of preparedness for potential risks and challenges. It can also lead to biased or flawed decision-making processes, as assumptions and estimates may not reflect the true nature of the situation. In other words, without accounting for uncertainty, the accuracy and reliability of your estimates – and the likelihood of the project's success – may be undermined.

4.14.3 Two types of uncertainty

Earlier, we mentioned two types of uncertainty, aleatory and epistemic.

The concepts of aleatory and epistemic uncertainty[39] apply in the context of M&V, as they can help to identify and manage different sources of uncertainty in estimates.

- *Aleatory uncertainty* refers to the inherent variability and randomness that is present in natural processes and events. It is the kind of uncertainty that cannot be reduced through additional information or analysis, as it is an inherent characteristic of the system itself. An example would be the probability distribution of measured wattage values from repeated tests on a piece of electrical equipment.
- *Epistemic uncertainty*[40] refers to uncertainty that arises from a lack of knowledge or information about a system or process. In the context of M&V, epistemic uncertainty can arise from factors such as incomplete or inaccurate data, measurement errors, modeling assumptions, and potential variations to baseline conditions. Epistemic uncertainty can be reduced through improved data collection and analysis, calibration

[39] https://www.sciencedirect.com/topics/engineering/epistemic-uncertainty

[40] https://www.hindawi.com/journals/jqre/2012/617481/#

of measurement equipment, and increased understanding of the system being measured. But ultimately, since it includes all the uncertainties around potential events in the reporting period, epistemic uncertainty is always lurking.

In the M&V conceptual framework, aleatory and epistemic uncertainty can be accounted for in different ways. Aleatory uncertainty is often quantified through statistical data analysis, but it cannot be eliminated entirely.

Epistemic uncertainty is inherently unknowable, but it can be simulated through sensitivity analyses, where different assumptions or scenarios are tested to understand their impact on the estimate.

By considering both aleatory and epistemic uncertainty in the M&V process, it is possible to enhance communication and understanding of the energy impact estimate – and the limitations of M&V.

4.14.4 Risk

We use statistical methods to arrive at quantitative estimates of uncertainty. As M&V professionals, we use uncertainty analysis to gauge our confidence in the results and to manage costs and expectations of the measurement approach. These can be seen as independent of the concept of responsibility for a particular outcome. Sometimes we just want to know roughly how much trust we have in our reported values. But, especially in performance contracting, there is another dimension to uncertainty. This is the concept of risk.

- *Risk* is the potential for loss or harm associated with a particular decision or course of action. It is a measure of the likelihood and magnitude of a negative outcome. Risk management is the process of identifying potential risks and developing strategies to mitigate or minimize them.

In the context of M&V, risk is most closely aligned with the process of settlement – the point where the purchaser of "savings" pays the provider. One can easily imagine that situations with minimal uncertainty can lead to confidence and trust, leading to a satisfying transaction for both parties (purchasing fuel at a gas station). However, if the data supporting the M&V results cannot support a high level of confidence, both parties might express concern, disappointment, or even outright disagreement. When money is on the line, both parties need to be prepared to deal with uncertainty before the settlement date.

4.14.5 More terminology

- *Confidence (level)* is the degree of certainty that can be placed in the accuracy or reliability of a particular measurement or estimate. It is often expressed as a percentage and indicates the range of values within which a specific measurement or estimate is likely to fall. Confidence intervals are commonly used in statistical analysis to describe the precision of a measurement or estimate. For example, a 95% confidence interval means that 95 out of 100 times the true value will fall between the upper and lower values specified by the confidence interval.
- *Probability* is the likelihood or chance of a particular outcome or event. It is a measure of uncertainty and is expressed as a number between 0 and 1, with 0 indicating that the outcome is impossible and 1 indicating that the outcome is certain. Probability is often used in statistical analysis to describe the likelihood of different events.
- A *parameter* is a fixed numerical value that characterizes a population or a model, such as the mean or standard deviation.
- A *coefficient* is a value that quantifies the relationship between two or more variables in a mathematical equation or model – for example, the slope of a linear regression line.
- A *variable* is a quantity that can take on different values in a given context and is often used to represent some property or characteristic of a system or phenomenon. Variables can be either independent, meaning their values can be manipulated or controlled, or dependent, meaning that other variables determine their values.

4.14.6 Error

GUM Definitions

The internationally recognized "Guide to Uncertainty in Measurements" (GUM) maintains a list of general terms.

The "Guide to the Expression of Uncertainty in Measurement" (GUM)[41] is a widely recognized international standard that provides guidelines for evaluating and expressing uncertainty in measurement results. According to the GUM, the term "error" is used in a specific context that differs somewhat from its colloquial usage. In the GUM, "error" is defined as follows:

[41] https://www.iso.org/sites/JCGM/GUM/JCGM100/C045315e-html/C045315e_FILES/MAIN_C045315e/03_e.html

Error (in the GUM):

- The GUM defines error as "the result of a measurement minus a true value of the measurand."

In this context, "measurand" refers to the quantity being measured, and the "true value" refers to the actual value of the quantity being measured. The GUM focuses on expressing uncertainty in terms of errors to assess the discrepancy between the measurement result and the true value.

It is important to note that the GUM uses the term "error" in this specific sense to discuss the difference between a measured value and the true value of a quantity. This differs from the broader use of the term "error" in everyday language, which can refer to mistakes, inaccuracies, or deviations from a desired outcome.

Ruling brings beer row to a head:

When is a pint not a pint? When it contains a frothy head of more than one fluid ounce, according to a ruling by consumer minister Melanie Johnson, which infuriated real ale enthusiasts yesterday.

Answering a parliamentary question, Ms. Johnson said the government would introduce legislation requiring a pint of beer to contain 95% liquid beer.

She was immediately accused of betrayal by campaigners, who said she had broken a labor promise to introduce a 100% liquid pint.

Pub companies have lobbied against tight legislation, claiming a 100% ruling would require millions of new glasses. But the Campaign for Real Ale has long opposed "legalised short measures," insisting any head should be above and beyond a pint.

4.14.7 Accuracy and precision

The GUM defines accuracy and precision in the context of measurement as follows:

- **Accuracy:**

Accuracy refers to the closeness of a measured value to the true value of the quantity being measured.

It is an indication of how well a measurement reflects the actual value of the quantity.

Accuracy takes into account both systematic errors (biases) and random errors (variations).

An accurate measurement is one that has a small systematic error and is close to the true value.

- **Precision:**

Precision refers to the closeness of agreement between individual measured values obtained from repeated measurements of the same quantity.

It reflects the degree of scatter or variability among the measured values.

Precision is related to the random errors in a measurement process.

A precise measurement is one where the measured values cluster closely together, even if they are not necessarily close to the true value.

In summary, accuracy is about how close a measurement is to the true value, while precision is about the consistency and reproducibility of measurements. A measurement can be accurate but not precise if it consistently deviates from the true value by a certain amount. Conversely, a measurement can be precise but not accurate if it consistently produces the same value but that value is far from the true value. The ideal measurement is both accurate (close to the true value) and precise (consistent and reproducible).

4.14.8 Bias

In the GUM, bias is defined as follows:

Bias: The "Guide to the Expression of Uncertainty in Measurement" (GUM) defines bias as "the result of a systematic error." In other words, bias refers to a consistent or systematic discrepancy between the average of measurement results and the true value.

Systematic errors are errors that consistently affect measurements in the same direction. These errors can arise from various sources, such as calibration issues, instrumentation limitations, or environmental conditions. Unlike random errors, which fluctuate around the true value, systematic errors consistently shift measurements in a specific direction.

Managing and correcting for bias is crucial in accurate measurement and estimation of uncertainties. Addressing systematic errors involves identifying their sources, quantifying their effects, and applying corrections or adjustments to measurement results to minimize their impact on the final outcome. It is important to identify and correct for bias in M&V to ensure that estimates are as accurate as possible. To address bias in M&V, you should use high-quality measurement and verification techniques designed to minimize measurement errors and account for any sources of bias in the data. The GUM provides guidance on how to account for bias and other sources of uncertainty when evaluating and expressing measurement uncertainties.[42]

4.14.9 Goodness of fit (model "quality")

Goodness of fit refers to how well a statistical model fits the observed data. In other words, it measures how well the model captures the patterns and variability present in the data.

Several measures of goodness of fit can be used to evaluate the performance of a model. Common measures include:

- *R-squared* measures the proportion of the variance in the dependent variable explained by a model's independent variables. A higher R-squared indicates a better fit.
- *Mean squared error (MSE)* measures the average squared difference between a model's predicted values and the data's actual values. A lower MSE indicates a better fit.
- *Root mean squared error (RMSE)* is the square root of the MSE. It is an often-used measure of the average error in a model's predictions.
- *Residual plots* show the differences between the predicted and actual values in the data. A good model will have residuals randomly scattered around zero, indicating that the model is capturing the patterns and variability in the data.

R-Squared

R-squared (also known as the coefficient of determination) is a measure of how well a statistical model fits the data. It represents the proportion of the total variation in the dependent variable (the variable being predicted) that can be explained by the independent variable(s) (the variables used to predict the dependent variable).

In simpler terms, R-squared measures how much of the variation in the data is accounted for by the model. An R-squared value of 1 means that the model explains all the variation in the data, while an R-squared value of 0 means that the model explains none of the variation in the data.

However, it is important to note that a high R-squared value does not necessarily mean the model is good or can accurately predict future observations.

---

[42] GUM Developing and Using Measurement Models = Page 59 https://www.bipm.org/documents/20126/50065290/JCGM_GUM_6_2020.pdf/d4e77d99-3870-0908-ff37-c1b6a230a337

It is just one measure of goodness of fit and should be used in conjunction with other measures and considerations.

CV(RMSE)

The coefficient of variation of RMSE, CV(RMSE), is a measure that combines the concepts of variability (as measured by the coefficient of variation or CV) and accuracy (as measured by the root mean squared error or RMSE). It is calculated as the ratio of the RMSE to the mean of the data, expressed as a percentage.

CV(RMSE) is often used to compare the performance of different models in predicting a variable. A lower CV(RMSE) indicates higher accuracy and lower variability in the predictions.

CV(RMSE) should be interpreted carefully, as it may not always be the most appropriate measure of model performance, depending on the specific application and goals. In some cases, other measures, such as R-squared, may be more appropriate.

Combining R-Squared and CV(RMSE)

Because R-squared and CV(RMSE) offer different perspectives on the accuracy and precision of a statistical model, considering them together can help provide a more comprehensive picture. You can also use these measures together with other diagnostic tools and methods. For example, visual inspection of residual plots can help to identify potential issues with model assumptions or data quality.

Residual Analysis

Residual analysis is a method used to evaluate the goodness of fit of a statistical model by examining the differences between the observed values and the predicted values (residuals). These residuals represent the deviation between the actual data points and the model's estimated values, and residual analysis involves assessing the distribution, pattern, and magnitude of these residuals to determine whether the model adequately captures the underlying data patterns. For example, if the residuals exhibit a non-random pattern or a systematic trend, this may suggest that the model is not adequately capturing all relevant variables or that underlying structural changes in the building were not accounted for in the model.

LINE

LINE analysis (*l*inear regression *in*teractive *e*stimation) is a statistical technique for analyzing and interpreting linear regression models. It entails manipulating the input variables of a linear model and examining the resulting changes in the output to assess the sensitivity and significance of the model's coefficients.

In M&V, LINE analysis helps us evaluate the accuracy and robustness of our regression models, assisting us in identifying which input variables have the greatest impact on energy consumption.

LINE analysis can even be used to explore potential scenarios through the manipulation of the model's input variables. For example, you could use LINE analysis to estimate the potential impact of different energy conservation measures or evaluate the effect of changing weather patterns on energy consumption.

Rounding

Rounding affects the accuracy of the reported value. When rounding a value, we need to consider the uncertainty associated with that value.

For example, if a value is reported as 10.1 with an uncertainty of 0.2, rounding to one decimal place would return a value of 10.1 ± 0.2, which could be reported as 10.1 ± 0.2 or 10.0 ± 0.2, depending on the rounding convention used.

It is generally recommended to use the "half-up" rounding convention, also known as "conventional rounding," where values that are halfway between two possible rounded values are rounded up. This helps minimize rounding bias and provides a consistent and predictable rounding method.

Reporting excess significant digits can create a false sense of precision or accuracy in the results being reported. For example, if a measurement is reported as 12.3456789, it implies that the value has been measured with extreme precision, to the ninth decimal place. However, in reality, it is highly unlikely that the measurement is that precise, and it may be subject to significant uncertainty or error.

In scientific and engineering contexts, it is important to report measurements and results with the appropriate level of precision, taking into account the uncertainty associated with the measurement or calculation. This can be achieved by rounding the reported values to an appropriate number of significant digits based on the level of uncertainty and the precision of the measurement instrument or calculation method.

Reporting results with the appropriate level of precision, including appropriate rounding, helps avoid false precision and accurately communicates the level of uncertainty or error associated with the results. This is important for maintaining credibility and stakeholder trust.

Example

Suppose an M&V analysis is conducted on a lighting retrofit project, and the project's impact is calculated to be a reduced consumption of 145,812 kWh per year with an uncertainty of ±5% at 95% confidence level. Reporting the impact as a reduction of 145,812.000 kWh per year implies a false precision unjustified by the level of certainty (145,812 $\sim$ 0.0007%). The true impact could be anywhere between 138,521.4 kWh per year and 153,102.6 kWh per year, at 95% confidence level.

Therefore, a more appropriate way to report the impact of the energy management activities might be as 146,000 kWh per year with an uncertainty of ±7300 kWh per year. This properly reflects the uncertainty in the estimate and avoids the false precision implied by reporting excessive significant digits.

4.14.10 Terminology time: "robust"

The word "robust" can have multiple meanings depending on the context in which it is used. In general, the term refers to the ability of a system or process to resist or tolerate variations, changes, or perturbations in its environment, inputs, or outputs without significant deterioration in performance.

In M&V, "robust" often refers to an M&V system's ability to produce accurate and reliable results despite uncertainties or variations in the measured data or assumptions.

However, "robust" is often used vaguely or ambiguously, without a clear definition or criteria for what constitutes robustness. For example, a product or system may be marketed as "robust" without any specific explanation of what that means in terms of performance or reliability.

In general, it is important to clarify the meaning and context of terms like "robust" to avoid confusion or misunderstandings. In the case of M&V, it is important to develop clear and well-defined criteria for what constitutes a robust M&V system and to evaluate different approaches based on their ability to meet those criteria.

4.14.11 Statistics and judgment

Statistics is a powerful tool for analyzing and interpreting data in M&V, but it is not a one-size-fits-all solution. As an M&V professional, you must consider the project's specific circumstances and make judgments about which statistical methods are most appropriate for the particular situation.

You must also exercise judgment[43] in determining the sample size and sample selection criteria, as well as in the interpretation of the statistical results. This requires a deep understanding of statistical concepts and knowledge of the underlying physics and engineering principles related to the energy consumption of the systems being evaluated.

In sum, while statistics is an important tool for M&V, it is not a substitute for professional judgment. M&V professionals must think about the context and employ their expertise to develop an appropriate M&V plan that accurately captures both the impacts of the energy management activities and their associated uncertainties.

4.14.12 Conclusion: conveying uncertainty

Teaching someone to appreciate the role of uncertainty requires a solid understanding of the concepts of uncertainty and how it manifests in M&V. As M&V professionals, we need to communicate these concepts clearly and accurately to diverse sets of stakeholders who may come in using different terminology.

One approach to securing alignment is to start with a basic overview of uncertainty in the context of M&V, including the different types of uncertainty that may arise (aleatory vs. epistemic, for instance) and how they can affect the accuracy and precision of M&V results. From there, you can introduce different statistical tools and techniques used to quantify and manage uncertainty, such as regression, residual, and LINE analysis.

Because stakeholders have backgrounds in different disciplines, you should try to use plain language and concrete examples to help them understand the implications of uncertainty in M&V. For example, discussing the potential impacts of overdesigning a system to meet a factor of safety can help engineers appreciate the need to understand the concept of uncertainty in M&V.

A number of available resources can help stakeholders – and you – learn more about the role of uncertainty in M&V, including textbooks, online courses, and industry organizations like the Efficiency Valuation Organization.[44] These resources can provide a more in-depth look at the various statistical methods used in M&V and practical guidance for working with stakeholders from different disciplines. Ultimately, the key to success in

[43]https://www.hindawi.com/journals/jqre/2012/617481/

[44]EVO's "Uncertainty Analysis Guide" is due for release in early 2024.

M&V is developing a deep understanding of the uncertainties involved and the tools and techniques used to manage them.

Exercise

Discuss the concepts of *accuracy* and *precision* with your team. Arrive at a shared definition of each term.

4.15 Metering and Calibration of Instruments

This chapter focuses on metering and calibration, which are crucial topics for the M&V professional. Metering can range from simple to complex. A facility may have a single meter or multiple meters already on-site – and, of course, additional meters will need to be installed in many cases.

For M&V professionals working on energy-savings performance contracts, a firm grasp of metering and calibration techniques is essential. However, many utility programs depend exclusively on metered data, minimizing the need for on-site involvement. Still, throughout this chapter, we will discuss the importance of combining both approaches for accurate results and offer practical knowledge for M&V professionals navigating the world of metering and calibration.

4.15.1 Metrology

Metrology is the science of measurement, and in M&V, accurate measurements are critical to quantifying the impacts achieved by energy management activities. This requires that the measurements be performed using validated instruments and methods, with appropriate calibration and traceability to national and international measurement standards.

A broad metrological infrastructure – including standards and traceability, calibration laboratories, and proficiency testing programs – helps keep the measurements for M&V accurate, reliable, and consistent.

For instance, the International Organization for Standardization (ISO) provides a range of international standards related to metrology and M&V, including ISO 50001 on energy management and ISO 17024 on personnel certification in energy management and energy efficiency.

4.15.2 Calibration

Calibration ensures that the instruments used to measure energy consumption are accurate and reliable. If the instruments used for taking measurements are not calibrated, the measurements may be inaccurate, leading to incorrect results.

Calibration involves comparing the readings of an instrument against a standard of known accuracy. The instrument is then adjusted as necessary. This typically happens in a specialized calibration laboratory, which has the equipment and expertise to perform precise measurements and adjustments.

Calibration is especially important for complex and sophisticated instruments, such as data loggers, which measure energy consumption over extended periods.

4.15.3 Measurement uncertainty analysis

Measurement uncertainty analysis (MUA) is the process of identifying, quantifying, and combining the various sources of uncertainty in a measurement. Typically performed using statistical methods, it is an important part of the M&V process.

The goal of MUA is to estimate the range of values within which the true value of the measurement is likely to lie. This information can be used to assess the accuracy of the measurement and make informed decisions about the results.

Many different sources of uncertainty can affect a measurement. Some of the most common sources include:

- **Measurement bias:** This is the difference between the true value of the measurement and the average of the repeated measurements. Various factors – such as the calibration of the measuring instrument, the skill of the operator, and environmental conditions – can cause bias.
- **Random error:** This is the variation in the repeated measurements due to chance. Random error cannot be eliminated but can be reduced by taking multiple measurements and averaging the results.
- **Environmental factors:** These can include factors such as temperature, humidity, and air pressure. Environmental factors can affect the accuracy of a measurement by changing the properties of the material being measured or the performance of the measuring instrument.

MUA can be complex, but it is an essential part of the M&V process. By understanding the sources of uncertainty in a measurement, M&V professionals can make more informed decisions about the results and ensure that the data is reliable.

4.15.4 GUM

The ISO Uncertainty series, commonly known as the "GUM" (Guide to the Expression of Uncertainty in Measurement), is a set of guidelines for evaluating and expressing uncertainty in measurements. The GUM is a widely recognized international standard that provides a comprehensive uncertainty analysis framework. It consists of several parts, including:

- **ISO/IEC Guide 98-3:2008 – "Guide to the Expression of Uncertainty in Measurement":** This is the main part of the series and provides a general framework for evaluating and expressing uncertainty in measurements.
- **ISO/IEC Guide 98-1:1993 – "Guide to the Expression of Uncertainty in Measurement – Propagation of Distributions Using a Monte Carlo Method":** This part provides a method for evaluating uncertainty using a Monte Carlo simulation approach.
- **ISO/IEC Guide 98-2:1996 – "Guide to the Expression of Uncertainty in Measurement – Extension to Any Number of Output Quantities":** This part provides guidance on how to extend uncertainty analysis to multiple output quantities.
- **ISO/IEC Guide 98-4:2012 – "Guide to the Expression of Uncertainty in Measurement – Example of the Calculation of Uncertainty":** This part gives examples of how to apply the GUM guidelines in specific measurement scenarios.

The GUM guidelines are based on the principles of probability theory and provide a consistent and rigorous framework for uncertainty analysis. More importantly, the series supplies practitioners with a common language for communicating uncertainty, which ensures that uncertainty estimates are understood and can be compared across contexts.

4.15.5 Commonly measured variables in M&V activities

There are several commonly measured variables in M&V, depending on the specific energy system or process being evaluated. Some of the most common include:

- *Energy consumption* is the amount of energy used by a building, facility, or process over a period of time. It is typically measured in units of kilowatt-hours (kWh) or British thermal units (BTUs).

- *Power demand* is the amount of electrical power required by a facility, system, or process at a given time. It is typically measured in units of kilowatts (kW).
- *Temperature* is often measured as part of M&V, particularly in HVAC (heating, ventilation, and air conditioning) systems. It is typically measured in degrees Celsius or Fahrenheit.
- *Humidity* is another variable often measured in HVAC systems, typically as a percentage of the maximum moisture the air can hold at a given temperature.
- *Pressure* is a common variable in many industrial processes, particularly those that involve fluids or gases. It is typically measured in pounds per square inch (psi) or kilopascals (kPa).
- *Flow rate* is a measure of how much fluid or gas is moving through a system per unit of time. It is typically measured in liters per minute (lpm) or cubic feet per minute (cfm).
- *Light intensity* is often measured as part of M&V in lighting systems, particularly for commercial or industrial applications, and is typically measured in lux or foot-candles.

These variables are often measured using sensors and meters installed as part of the M&V process.

Many resources are available online for more information on measuring these variables. Here are some you might find helpful:

[List better resources.]

- **Energy consumption:** The U.S. Department of Energy's Energy Information Administration website has information on measuring energy consumption in facilities and industrial processes. The EIA also provides a list of energy metering equipment manufacturers and suppliers.
- **Power demand:** The National Renewable Energy Laboratory (NREL) has published a comprehensive guide on measuring and analyzing power demand.
- **Temperature and humidity:** The National Institute of Standards and Technology (NIST) provides detailed information on measuring temperature and humidity, including calibration procedures and uncertainty analysis.
- **Pressure:** The American Society of Mechanical Engineers (ASME) provides information on pressure measurement and instrumentation, including standards and guidelines.

- **Flow rate:** The International Organization for Standardization (ISO) has published a series of standards related to flow measurement, including ISO 5167 for orifice meters and ISO 4185 for turbine meters.
- **Light intensity:** The Illuminating Engineering Society (IES) provides guidelines and standards for measuring light intensity and other lighting-related variables.

In addition to these resources, manufacturers and suppliers of sensors and meters often provide detailed specifications and user manuals that can be useful for understanding how to measure specific variables.

4.15.6 Different types and uses of metering systems

Metering systems play a critical role in M&V by providing data on energy consumption, demand, and other key variables. Here are some common types of metering systems used in M&V, along with their uses:

- *Electrical submetering systems* measure the energy consumption of individual systems or components within a building or industrial process.
- *Energy management systems (EMS)* are software-based systems that integrate data from multiple sources, including metering systems, to provide real-time information on energy consumption, demand, and other key variables.
- *Building automation systems (BAS)* are similar to EMS but focus more on controlling and optimizing building systems such as HVAC, lighting, and security. They may include metering capabilities to provide real-time data on energy use.
- *Utility submetering systems* measure the energy consumption of individual tenants or departments within a building or complex.
- *Process metering systems* measure variables such as temperature, pressure, flow rate, and chemical composition in industrial processes. This can help to optimize process efficiency and reduce energy use.
- *Portable metering systems* temporarily monitor energy consumption or process variables during commissioning, troubleshooting, or other short-term activities.

In addition to these types of metering systems, many meters and sensors are available to measure specific variables such as temperature, pressure, flow rate, and light intensity.

Best Practices for Minimizing Uncertainty in Metering

Some best practices for reducing uncertainty in the measurement, logging, communication, storage, calculation, and stipulation of metering data include:

- Ensure all measurement equipment is calibrated and maintained regularly.
- Use high-quality sensors and meters with a high degree of accuracy and precision.
- Use automated meter reading (AMR) or advanced metering infrastructure (AMI) systems.
- Implement data validation and quality control processes to identify and correct any errors or outliers in the data.
- Use secure and reliable communication protocols to transmit data from the sensors or meters to the data storage and analysis system.
- Ensure that the data is stored in a secure and accessible manner, with appropriate backup and recovery processes in place.
- Use standardized calculation methods and algorithms to maintain consistency.
- Use internationally recognized standards for uncertainty analysis, such as the ISO GUM.
- Implement clear and consistent documentation and reporting procedures to ensure stakeholders can easily understand and use the data.

Safety in Metering

Safety comes first. Electrical and mechanical hazards can be present in metering systems, particularly in industrial or high-voltage applications. Proper installation, maintenance, and operation of metering systems are critical to the safety of personnel and equipment.

Cost

The cost of installing and maintaining metering systems can be considerable. However, the benefits of accurate and reliable metering data can outweigh these costs over time.

Tools of the Trade

Current Transformers

Current transformers (CTs) measure the electrical current flowing through a conductor. Here are some common types of CTs and their respective benefits:

- *Wound primary CTs* have a primary winding wrapped around the measured conductor and a secondary winding providing a reduced current output to the metering system. They are simple and robust and can be used for many current ratings.
- *Split-core CTs* have a split housing that can be opened to install or remove them from the measured conductor without cutting or disconnecting it. This can simplify installation and reduce downtime in retrofit applications.
- *Clamp-on CTs* do not require any physical connection to the measured conductor but instead use a magnetic field to induce a current in a secondary winding. This can simplify installation and eliminate the need for downtime, but it may be less accurate than other types of CTs.
- *Rogowski coils* are flexible coils wrapped around irregularly shaped conductors or cables. They are lightweight and easy to install and can be used for many current ratings.
- *Busbar CTs* are designed to be mounted directly onto a busbar, simplifying installation and eliminating the need for additional wiring. They can be used for a wide range of current ratings and are often used in high-voltage applications.
- *Bar-type CTs* have a rectangular shape that allows them to be mounted directly onto a busbar or other conductor. They are simple and robust and can be used for many current ratings.

Power Meters

Power meters are electronic devices that measure various electrical parameters such as voltage, current, power, and energy consumption. Common types of power meters include:

- *Digital multimeters (DMM)* are handheld instruments that can measure various electrical parameters such as voltage, current, resistance, and continuity. They are portable, easy to use, and can provide accurate measurements for low-power applications.
- *Clamp meters* are handheld instruments that measure current flowing through a conductor without disconnecting it. They are portable and easy to use and can provide accurate measurements for medium- to high-current applications.
- *Panel meters* can be mounted on a control panel or equipment to display electrical parameters such as voltage, current, power, and energy consumption. They are easy to read and can provide real-time information.

- *Smart meters* can communicate with other devices and systems to provide real-time information on energy consumption and demand. They can have advanced features such as two-way communication, remote monitoring, and control and can be integrated with other smart building systems.
- *Power quality analyzers* measure various electrical parameters such as voltage, current, power, and energy consumption and identify power quality issues such as harmonics, voltage fluctuations, and transients. They are useful for troubleshooting electrical issues and ensuring power quality for sensitive equipment.
- *Energy management systems* are comprehensive systems that monitor, control, and optimize energy consumption and demand in a facility or system. They can integrate with various facility components such as lighting, HVAC, and equipment and provide advanced features such as real-time energy monitoring, demand response, and energy conservation measures.

Temperature Sensors

Temperature sensors are electronic devices that measure temperature, an important parameter in many applications. Here are some common types of temperature sensors and their respective benefits:

- *Thermocouples* consist of two wires of different metals joined together at one end. The temperature difference between the two ends of the wires creates a voltage, which can be measured to determine the temperature. Thermocouples are durable, can operate over a wide temperature range, and have a fast response time.
- *Resistance temperature detectors (RTDs)* consist of a thin wire made of a metal such as platinum, which changes resistance as the temperature changes. RTDs are highly accurate, stable, and have a linear response to temperature changes.
- *Thermistors* consist of a ceramic or polymer material resistor that changes resistance as the temperature changes. Thermistors are highly sensitive to temperature changes, have a fast response time, and are relatively inexpensive.
- *Infrared (IR) temperature sensors* are non-contact sensors that measure the infrared radiation emitted by an object to determine its temperature. IR temperature sensors are fast, accurate, and can measure temperature without coming into contact with the measured object.

- *Fiber optic temperature sensors* use optical fibers to transmit temperature measurements. They can be used in harsh environments such as high-pressure, high-voltage, or hazardous areas and provide accurate measurements over a wide temperature range.

Lighting Loggers

Lighting loggers measure lighting conditions, such as light intensity, luminance, and color temperature. Here are some common types:

- *Lux meters* measure the intensity of light in lux or foot-candles. They are easy to use, affordable, and provide accurate light intensity measurements.
- *Spectrophotometers* measure the color of light and can provide detailed information about the spectral composition of light. They are commonly used in colorimetry and photography.
- *Data loggers* can be programmed to record light measurements over a period of time and can provide data for analysis and optimization of lighting systems.
- *Colorimeters* measure the color of light and can provide accurate measurements of color temperature and color rendering index (CRI). They are commonly used in photography, graphic design, and lighting design.
- *Photometers* measure the brightness of light sources and can provide accurate measurements of luminance and illuminance. They are commonly used in lighting design, photography, and film production.

Flow Meters

Flow meters measure the flow rate of liquids or gases. Here are some common types of flow meters and their respective benefits:

- *Differential pressure flow meters* measure the pressure drop across an obstruction in the flow path to determine the flow rate. They are easy to use, cost-effective, and provide accurate measurements for various fluid types and flow rates.
- *Turbine flow meters* employ a turbine to measure the velocity of a fluid, which is then used to calculate flow rate. They are highly accurate, have a wide measurement range, and can be used for both liquid and gas applications.
- *Ultrasonic flow meters* use ultrasound technology to measure the velocity of the fluid, which is then used to calculate the flow rate. They are

non-intrusive, easy to install, and can be used for both liquid and gas applications.

- *Magnetic flow meters* use a magnetic field to measure the flow rate of conductive fluids. They are highly accurate, have no moving parts, and can be used for a wide range of fluid types and flow rates.
- *Vortex flow meters* use the Kármán vortex street principle to measure the flow rate of liquids and gases. They are highly accurate, have no moving parts, and can be used for a wide range of fluid types and flow rates.

4.15.7 Power factor – real power – other metering and measurement concepts

Note: "Electrical Engineering for Non-Electrical Engineers" by S. Bobby Rauf is a great reference for understanding electrical systems.[45]

Power factor is a measure of the efficiency of electrical power usage in a system. In M&V, power factor measurement is important for accurately determining the true power consumption of a system, as opposed to the apparent power consumption measured by traditional power meters. This is because power factor represents the ratio of real power (in watts) to apparent power (in volt-amperes) in an AC electrical system. A low power factor indicates that the system is not efficiently using the electrical power supplied to it, resulting in higher energy costs and potential equipment damage.

Several tools can measure power factor, including:

- **Power factor meters:** These devices measure the power factor of an AC electrical system in real time. They are available as standalone meters or can be integrated into power quality analyzers.
- **Digital multimeters:** Some high-end digital multimeters can measure power factor in addition to voltage, current, and resistance.
- **Power quality analyzers:** These devices can measure various parameters related to power quality, including power factor. They are useful for diagnosing power quality issues and identifying energy-saving opportunities.
- **Clamp meters:** Some clamp meters can measure power factor in addition to current and voltage. They are useful for quickly measuring power factor in a specific circuit or piece of equipment.

---

[45]https://www.riverpublishers.com/book_details.php?book_id=901

The common issues with metered data are cost, accuracy, and precision specifications (confidence interval and confidence level).

If accuracy is reported as an interval, it is likely referring to the confidence interval, which represents the range within which the true value is expected to fall with a certain level of confidence. The confidence level is typically also reported alongside the confidence interval. For example, if a meter is rated as having an accuracy of $\pm 1\%$ with a 95% confidence level, there is a 95% chance that the true value falls within the range of $\pm 1\%$ of the measured value.

If the confidence level is not explicitly stated, it may be necessary to contact the manufacturer or supplier of the meter to obtain this information. It is also possible to estimate the confidence level based on the reported accuracy and the statistical method used to calculate the confidence interval, but this would require knowledge of the specific measurement and statistical techniques involved.

4.15.8 Energy information systems[46]

Energy information systems include the following. Note that other systems may be relevant as well, as the field of energy management and monitoring is constantly evolving.

- Energy management information system (EMIS)
- Building automation system (BAS)
- Advanced metering infrastructure (AMI)
- Supervisory control and data acquisition system (SCADA)
- Demand response automation server (DRAS)
- Enterprise energy management system (EEMS)
- Building energy information system (BEIS)
- Facility smart grid information model (FSGIM)
- Automated fault detection and diagnostics (AFDD)
- Virtual energy audit tool (VEAT)
- Green button connect (GBC)
- Energy performance contracting (EPC) platform

BMCS

A building management and control system (BMCS) is a computer-based control system that manages the energy-consuming systems in a facility, such

[46]https://elibrary.riverpublishers.com/ebook_read.php?id=806, https://elibrary.riverpublishers.com/ebook_details.php?id=795

as HVAC, lighting, and other energy-consuming equipment. It can be used in M&V to collect, monitor, and analyze data on energy use and identify opportunities for energy management activities. Here are three examples of how a BMCS can be used in M&V.

- **Baseline energy consumption:** A BMCS can be used to establish a baseline of energy consumption for a building by monitoring the energy consumption of each system for a set period, such as a month or a year.
- **Real-time monitoring:** A BMCS can be used to monitor the energy consumption of each system in real time. This can help M&V professionals, energy auditors, and facility operators identify issues like equipment malfunctions or abnormal energy use patterns. Facility operators can take corrective action quickly and minimize wasted energy.
- **Energy savings:** A BMCS can be used to implement energy conservation measures in a building. For example, the BMCS can automatically adjust the temperature in a room based on occupancy or the lighting levels based on natural light levels.

4.15.9 Digital twins

A digital twin is a virtual representation of a physical system that can be used to simulate the behavior of the asset or system, predict its performance, and identify potential problems before they occur. Energy information systems such as BAS, AMI, EMIS, and SCADA can provide real-time data on energy consumption and facility operations, which can be used to develop digital twins for M&V and other energy management purposes. Overall, the use of energy information systems in combination with digital twins can provide a powerful tool for improving energy efficiency and reducing operating costs in facilities.

There are several resources available to keep track of the digital twin industry, including the following:

- **Digital Twin Consortium[47]:** A global organization that promotes digital twin technology across various industries.
- **Industrial Internet Consortium[48]:** An organization focused on advancing the adoption of the Industrial Internet of Things (IIoT), which includes digital twin technology.

[47] https://www.digitaltwinconsortium.org
[48] https://www.iiconsortium.org

- **Gartner[49]:** A research and advisory firm that provides insights and analysis on emerging technology trends, including digital twin technology.
- **ARC Advisory Group[50]:** A research and advisory firm that provides market research and analysis on the industrial automation industry, including digital twin technology.

Attending industry conferences and events focused on digital twin technology can be a great way to stay up to date on the latest trends and developments in the industry. Some examples include the annual Digital Twin Forum and the Digital Twin Conference and Expo.

4.16 Field Work: Formal Accreditation, Safety, OSHA, and NFPA

Safety regulations and procedures must be followed when working with electrical equipment, including when opening electrical panels.

In the United States, the Occupational Safety and Health Administration (OSHA) has specific requirements for working with electrical equipment under its Electrical Safety Standard (29 CFR 1910.331-335). This standard outlines requirements for the installation, use, and maintenance of electrical equipment and the training and qualifications of personnel who work with this equipment.

In addition, the National Electrical Code (NEC), published by the National Fire Protection Association (NFPA), provides guidelines and standards for the safe installation of electrical equipment and systems.

Field technicians who open electrical panels must be trained in electrical safety and follow established safety procedures to minimize the risk of electric shock, arc flash, and other electrical hazards. They may also be required to wear personal protective equipment such as gloves, safety glasses, and arc flash suits, depending on the specific hazards and risks associated with the electrical equipment being worked on.

The specific safety regulations and procedures for working with electrical equipment may vary depending on the location and jurisdiction; so you should consult with local authorities and industry standards to ensure compliance.

For other systems that may be encountered in M&V, such as flow measurements in a chiller plant, the qualifications for installing metering

---

[49]https://www.gartner.com/en/information-technology/glossary/digital-twin

[50]https://www.arcweb.com

equipment may depend on several factors, including the specific equipment being installed, the complexity of the system, and the local regulations and standards.

Generally, it is important to ensure that the individuals installing metering equipment have the necessary technical knowledge and experience to do so safely and effectively. This may include training in the specific type of metering equipment being installed and knowledge of the underlying principles of measurement and control.

Finally, depending on the specific application, other requirements or regulations may need to be considered when installing metering equipment. To return to our previous example, in some cases, the installation of flow meters in chiller plants may be subject to water treatment and environmental protection regulations.

4.16.1 Incident reporting

Various incident reporting forms are available, depending on the industry and the type of incident being reported. Some common incident reporting forms include:

- **Workplace Incident Report Form[51]:** This is a general incident reporting form used by employers to document and investigate work-related accidents, injuries, and illnesses. It usually asks for information about the employee, the nature of the incident, the location and time of the incident, and any witnesses or contributing factors.
- **Vehicle Accident Report Form:** This form documents vehicle accidents that occur on the job, such as accidents involving company-owned vehicles or personal vehicles driven for work purposes.
- **Environmental Incident Report Form:** This form is for reporting incidents that result in environmental damage or pollution, such as spills, leaks, or releases of hazardous materials.
- **Product Safety Incident Report Form:** This form is used to report incidents involving defective or unsafe products, such as injuries or deaths caused by faulty equipment or consumer products.

It is important to use the appropriate incident reporting form for the specific type of incident being reported and to ensure that all required information is included to facilitate a thorough investigation and prevent future incidents.[52]

[51] https://www.osha.gov/sites/default/files/publications/osha3071.pdf

[52] https://www.osha.gov/ords/imis/accidentsearch.accident_detail?id=201169307

4.17 Availability of Data

There are various examples of how data availability can be an issue for M&V, including:

- **Lack of metering infrastructure:** Often, facilities do not have adequate metering infrastructure in place to measure the energy consumption of individual systems or equipment. This makes it difficult to isolate the energy consumption of specific systems, which is necessary for accurate M&V.
- **Data inaccessibility:** Even if metering infrastructure is in place, data may not be easily accessible for various reasons, such as lack of communication protocols, meter incompatibilities, or data in different formats. This can lead to delays or inaccuracies in M&V analysis.
- **Incomplete or inaccurate data:** Even when available, data may be incomplete or inaccurate, affecting the accuracy of M&V analysis. For example, missing data from a meter for a particular period can make it difficult to accurately determine energy consumption.
- **Data quality issues:** Data quality can be affected by factors such as meter calibration issues, incorrect settings or configurations, or data transmission errors.
- **Limited historical data:** Historical data is essential for accurate M&V analysis, but, often, data is only available for a limited period. This can make it challenging to determine energy consumption trends over time and accurately assess the impact of energy management activities.
- **Privacy concerns:** In some cases, data may be inaccessible due to privacy concerns, such as when metering data contains sensitive information that cannot be shared with third parties.

To mitigate the challenges posed by data availability issues, follow these best practices:

- Before conducting M&V, establish clear data requirements upfront and communicate them to stakeholders.
- Develop contingency plans in case necessary data is unavailable. This could include finding alternative data sources, using proxies or estimates, or adjusting the scope of the analysis.
- Ensure that the data collected is of high quality by conducting quality control checks, verifying data accuracy, and addressing missing or erroneous data.

- Utilize appropriate data analysis techniques for the data available and the objectives of the analysis.
- Consider the use of advanced data analytics such as machine learning and artificial intelligence when data is limited or difficult to obtain. These techniques can help to uncover patterns and relationships in the data that may not be apparent using traditional analysis methods.
- Collaborate with stakeholders to understand their needs and goals and ensure that the M&V process is aligned with these objectives. This helps ensure the necessary data is collected and that the analysis focuses on the most important issues.

4.17.1 Data access issues

The types of data needed for M&V vary depending on the project and the goals of the M&V program, but, in general, the following types of data are commonly used:

- *Energy consumption data* includes the amount of energy used by equipment, facilities, or systems.
- *Weather data* includes information on temperature, humidity, and other weather conditions.
- *Operational data* describes how facilities and equipment are used, such as occupancy patterns and operating schedules.
- *Facility characteristics data* includes facility size, construction type, and other relevant physical characteristics.
- *Financial data* includes data on project costs and targets, which are used to calculate return on investment and other financial metrics.

Accessing the data needed for M&V can be a challenge for several reasons:

- **Data availability:** In some cases, the data needed for M&V may not be readily available, particularly if it is stored in different systems or owned by different stakeholders.
- **Data quality:** Even when data is available, it may not be of sufficient quality to support accurate M&V. This can be due to errors in data collection or processing or inconsistencies in how data is reported.
- **Data integration:** M&V often requires combining data from multiple sources, which can be challenging if the data is stored in different formats or uses different units of measurement.
- **Privacy and security:** Data used in M&V may be subject to privacy and security restrictions, particularly if it includes information on individual behavior or usage patterns.

- **Cost:** Collecting and analyzing data for M&V can be expensive, particularly for large or complex projects.

To address these challenges, you should work closely with the project team and other stakeholders to ensure that the necessary data is available and is of sufficient quality.

4.17.2 Data is always lost

Missing data is a common challenge in M&V. Here are some strategies for handling this problem:

- **Imputation:** One strategy for handling missing data is to fill in the missing values using a statistical imputation method. This involves estimating the missing data based on the observed data and statistical models. Various imputation methods are available, including mean imputation, regression imputation, and multiple imputation.
- **Substitution:** Another strategy is to substitute the missing data with similar data from another source. For example, if data is missing for a specific time period, one might use data from a similar period in a different year as a substitute.
- **Exclusion:** In some cases, missing data cannot be imputed or substituted, and exclusion may be necessary. However, be careful about which data points to exclude, as this can introduce bias into the analysis. Exclusion should only be done if there is a valid reason for doing so, such as if the data is clearly erroneous or if the missing data is minimal.
- **Sensitivity analysis:** Finally, you can perform a sensitivity analysis to assess how the missing data might affect the results. This involves testing the robustness of the results to various assumptions about the missing data.

Generally, the best strategy for handling missing data will depend on the context and the available data.

Several specific issues can prevent data access for M&V professionals. Some examples include:

- **Firewalls:** Data may be stored on servers behind firewalls that restrict access to outside users.
- **Lack of equipment:** In some cases, data may not be available because the necessary equipment to collect it was not installed or functioning properly.

- **Data storage:** Data may be stored in a format that is difficult to access or analyze.
- **Privacy concerns:** Data may contain sensitive information that cannot be shared with outside parties.

To ensure the best results in these situations, you should work closely with stakeholders to understand the specific challenges and find solutions acceptable to all parties involved. Additionally, you should maintain clear communication with stakeholders throughout the process so that expectations are met and any issues are addressed promptly.

Example: U.S. Military Bases

The U.S. Department of Defense (DOD) presents a particularly good example of how M&V can be challenging due to data access issues. The DOD has a vast and complex infrastructure that includes many different types of buildings, facilities, and equipment. It also has unique energy and sustainability requirements, including mandates to reduce energy consumption and increase the use of renewable energy sources. However, much of the DOD's data is classified or sensitive, making it difficult to access and use for M&V purposes.

As a result, the DOD has launched several initiatives to address data access and other energy management challenges. For example, the DOD has established energy management systems and data analytics tools to help collect and analyze data on energy use and identify areas for improvement. It has also worked with industry partners to develop new technologies and approaches for M&V that can be used across its facilities and operations. Specific examples of DOD energy management partnerships and initiatives include:

- The U.S. Army Engineering and Support Center, in collaboration with the National Renewable Energy Laboratory, developed the ESPC ENABLE program. This program provides a streamlined process for small federal facilities to implement energy efficiency upgrades and improve their energy management practices.
- The Environmental Security Technology Certification Program (ESTCP) funds and demonstrates innovative technologies that improve the environmental performance of DOD facilities and operations. This includes M&V technologies and approaches, such as automated fault detection and diagnostics, energy modeling, and real-time energy monitoring.

- The Energy Conservation Investment Program (ECIP) funds energy efficiency projects at DOD facilities, including M&V measures and technologies. ECIP projects must demonstrate a positive return on investment and meet specific energy impact and performance criteria.
- The Energy Data Management (EDM) program provides a centralized platform for collecting, analyzing, and reporting energy use data across DOD facilities and operations. This includes data on energy consumption, cost, and greenhouse gas emissions.
- The Advanced Metering Infrastructure (AMI) program installs advanced energy meters and data collection systems at DOD facilities to improve energy management, emphasizing real-time energy monitoring, analytics, and reporting.

4.18 Open-source Software

Open-source software offers access to powerful, highly adaptable data analysis and modeling tools. It sets users free from the constraints of proprietary software, as the open and freely available source code allows them to customize these tools to fit their specific needs.

In M&V, open-source software promotes transparency and standardization. As the underlying code is available for everyone to inspect, users can verify models' and calculations' accuracy and reliability. The open nature of the software also allows the identification and rectification of potential errors or biases. The spirit of collaboration inherent in open-source development allows the M&V community to establish common standards, best practices, and a shared knowledge base that benefits all.

Jupyter Notebook

Commonly used for Python, Jupyter Notebook supports multiple programming languages. Its name, derived from the languages that it supports, reflects this flexibility: Julia, Python, and R. Jupyter Notebooks allow users to write and execute code in a web-based environment and offers the functionality to create and share documents that contain live code, equations, visualizations, and narrative text.

The FLEX Market project exemplifies the use of Jupyter Notebook in M&V. This project aims to develop and test a platform for flexible demand response in the electricity grid. The project team has developed a Jupyter Notebook that offers insights into using machine learning algorithms to forecast energy demand and supply in real time, complete with detailed explanations and code examples.

CalTrack Methods[53]

OpenEE developed CalTRACK, an open-source tool that uses the Time-of-Week and Temperature (TOWT) model from the Lawrence Berkeley National Laboratory (LBNL). This model estimates a building's baseline energy consumption, which is essential for determining the impacts of efficiency measures.

CalTRACK's main objective is to standardize M&V methods and foster a culture of data sharing and best practices among stakeholders in the energy efficiency industry. The software is promoted by LF Energy, a Linux Foundation project that supports the development of open-source technologies in the energy sector.[54]

Energy Modeling Simulation Software

Open-source software is not limited to data analysis. Several building simulation software programs are available for free:

- **EnergyPlus:** Developed by the U.S. Department of Energy and used widely to simulate energy consumption in buildings.
- **OpenStudio:** From the National Renewable Energy Laboratory, it offers a user-friendly interface.
- **DesignBuilder:** Developed in the U.K. with an emphasis on sustainable design (based on Energy Plus, partially open source).
- **TRNSYS:** From the University of Wisconsin, it simulates energy systems and buildings.

DOE-2

DOE-2 is a building energy simulation software developed by the U.S. Department of Energy. It is freely available, and while it is not technically open source, its source code can be accessed upon request. Newer, more user-friendly tools with advanced features like EnergyPlus and OpenStudio have largely replaced it. However, its historical significance as a trailblazer in building energy performance simulation and evaluation of energy conservation measures is noteworthy.

[53]http://docs.caltrack.org/en/latest/methods.html
[54]http://docs.caltrack.org/en/latest/methods.html

ECAM
https://sbwconsulting.com/ecam/

Open-source Software for Statistics

There are several open-source statistical software options available as well, including the following:

- **R:** A free and open-source programming language for statistical computing and graphics. It enables a range of statistical and graphical techniques, including linear and nonlinear modeling, statistical tests, time-series analysis, and classification. R is widely used in academia, research, and industry. The Department of the Built Environment at the National University of Singapore maintains a great resource for using R in building energy simulation modeling.[55]
- **Python:** A high-level, general-purpose programming language that can be used for various purposes, including data analysis and statistical modeling. It has several powerful libraries, including NumPy, Pandas, and SciPy, which provide a range of functions for scientific computing, data analysis, and visualization.
- **OpenBUGS:** A free and open-source software package for Bayesian statistical analysis. It provides a flexible and extensible framework for building and running models using the BUGS language, a declarative language for specifying statistical models.
- **Stan:** An open-source probabilistic programming language for Bayesian inference. It provides a flexible and efficient framework for building and running models using a range of statistical algorithms, including Markov chain Monte Carlo (MCMC) and optimization methods.

4.19 Communication

By this point in the book, you likely recognize the importance of soft skills in M&V – effective communication, especially.

A complex M&V project requires various specialists working together as a team, including engineers, analysts, managers, and sometimes even legal and financial professionals. Each team member brings a unique skill set, contributing to the project's success. No one individual has the expertise necessary to drive the project alone.

[55] https://r4bes.com

Even in simpler projects, not all stakeholders have an engineering background. Facility owners, community representatives, and end users may need to understand the project's scope, process, and benefits. As M&V professionals, we are responsible for delivering information in an accessible and comprehensible manner, avoiding jargon as much as possible and using plain language to describe complex technical concepts. You should consider creating a glossary of terms or conducting a technical briefing before the project starts.

From the beginning, you should establish a clear and concise communication plan that outlines all team members' and stakeholders' goals, objectives, roles, responsibilities, and timelines. The plan should also include the communication channels to be used, frequency, and escalation procedures in case of issues.

Consider this sample communication plan.

Goal

- Ensure all stakeholders are informed about the M&V process and progress and facilitate collaboration and transparency.

Tasks

- Educate stakeholders about the M&V process, including technical terms and key concepts.
- Issue regular updates on progress and results to stakeholders.
- Facilitate collaboration and information-sharing among stakeholders.

Target Audience

- Energy managers
- Facilities managers
- Data analysts
- Project managers
- Building occupants
- Senior leadership

Messages

- Explanation of the M&V process and purpose
- Updates on progress and results
- Clarification of technical terms and concepts
- Reminders of stakeholder roles and responsibilities

Channels

- Regular meetings and check-ins (in-person or virtual)
- Email updates and newsletters
- Project status reports
- Presentations at team meetings or stakeholder events
- Project website or shared document platform

You should also use visual aids such as diagrams, charts, and graphs to illustrate technical concepts and depict the project visually. This can be especially helpful for non-technical stakeholders.

In lengthy, complex projects, some team members or stakeholders will likely come or go over the project's lifecycle. These changes can pose challenges from a project management perspective; however, you can smooth the transition and maintain project momentum by employing some of the strategies below:

- Facilitate a "warm" (detailed) handover when possible. This should include project documentation, technical specifications, and ongoing issues or challenges.
- Review the project with the new team members to ensure they fully grasp what they need to know.
- Provide ongoing training and support to all team members, especially new ones, so that they have the necessary skills and knowledge to carry out their roles effectively.
- Update the communication plan as needed to loop in the new team members and reflect any changes to communication channels or procedures.

4.19.1 Data formats

Generally speaking, M&V data appears in three formats, each giving a unique perspective (see Table 4.2):

- Tables
- Scatter plots (for relationships between variables)
- Time series (for data patterns over time)

Line charts, also known as line graphs, are commonly used for visualizing time-series data. In a line chart, data points are plotted on a graph with time on the horizontal x-axis and the data point's value on the vertical y-axis. A line connects the data points, revealing a trend or pattern over time, making line charts a useful tool for data analysis in fields such as finance, economics, and science.

Table 4.2 Data Format Strengths and Weaknesses.

Format	Strength	Weakness
Table	Easy to recognize values	Lacks the readability of a visualization
Scatter plot	Clear demonstration of relationship between independent (x) and dependent (y) variables	Provides no information on the timing of the data
Time-series graph	Clear demonstration of temporal patterns	Does not directly depict strength of relationships between x and y variables

When reporting, you should be able to select the most appropriate format for a given set of data and present the data clearly and concisely in a way easily understood by stakeholders. The table below lists the formats' strengths and weaknesses.

4.19.2 Data visualization

As an M&V professional, you are frequently required to "explain" M&V to people unfamiliar with the concept. While energy (and other) data are critical to performing M&V, raw numbers alone fail to convey the challenges of creating a baseline, estimating uncertainty, performing adjustments, and addressing other issues that arise.

One powerful tool for communicating M&V concepts is data visualization, the practice of using tools and techniques to represent data in visual formats, such as charts, graphs, and diagrams, to help people understand complex datasets more easily.

There are many great sources of tools for data visualization. Python Libraries for Time Series and Data Visualization:

If you prefer to work with code, Python libraries like Pandas, Matplotlib, Seaborn, and Plotly are excellent choices for creating time series and x/y chart visualizations. You can leverage these libraries to visualize energy data, perform data analysis, and build custom dashboards.

R Statistical Software: R is another powerful tool for data visualization and analysis.

4.20 Valuation

Assigning a value to the impacts of energy management activities can be challenging due to various factors. Some of these factors include:

- **Baseline uncertainty:** There may be uncertainty in defining the baseline energy usage and corresponding energy cost, especially if seasonal or weather-related variations affect energy usage.
- **Market uncertainty:** Energy prices fluctuate significantly over time.
- **Measurement and verification uncertainty:** Errors may arise during the M&V process.
- **Non-energy benefits:** Energy efficiency measures can also provide non-energy benefits, such as improved comfort, indoor air quality, and reduced maintenance costs. However, the value of these benefits may not be easily quantifiable.

However, certain practices can help you mitigate the challenges posed by these factors:

- Define a clear, consistent methodology for quantifying and assigning value to the impacts of energy management activities.
- Establish a well-defined baseline period and adjust for any seasonal or weather-related variations that may affect energy usage.
- Use market data and projections to estimate future energy prices and incorporate sensitivity analyses to account for uncertainty.
- Implement robust M&V procedures.
- Conduct comprehensive cost–benefit analyses that account for both energy and non-energy benefits.
- If possible, use standardized protocols or tools for cross-project consistency.
- Involve stakeholders in the valuation process and provide them with clear and transparent information about the methodologies and assumptions used.

4.20.1 Rates, data, and avoided costs

The complexity of rate structures can also create difficulties for valuation. In many cases, there are multiple rates for different time periods, different tiers of energy usage, and different types of energy (e.g., electricity vs. natural gas). Thus, to properly value the impacts of energy management activities, you may need to apply the marginal price to both the adjusted baseline and actual energy use.

For example, if a utility company employs time-of-use (TOU) rates, which vary depending on the time of day or day of the week, you would need to determine the energy usage during each time period and apply the corresponding TOU rate for an accurate valuation. If the utility has tiered rates, where the price per unit of energy increases as usage increases, you must also determine which tier the energy usage in question falls into and apply the appropriate rate for that tier.

As always, the right approach will depend on the context. Sometimes, it may be appropriate to use a different approach, such as a fixed price or a blended rate. Whatever the case, you should work closely with stakeholders to determine the best approach to valuation.

4.20.2 Savings load shapes

As discussed in Section 4.8.3, the shape of the load profile or energy use pattern before and after a retrofit can affect the value of the impact of an energy management activity. For example, suppose the retrofit reduces energy use during peak hours, when energy costs are the highest. In that case, the impact of the energy management activity may be more valuable than if the retrofit reduces energy use during off-peak hours – because the marginal cost of energy, or the cost of producing an additional unit of energy, is typically higher during peak hours.

When it comes to valuation, this so-called *savings load shape* can add another layer of complexity to the analysis. You might need to consider not only the marginal price of energy but also the time of day or season when the energy impact occurs. In some cases, applying different marginal rates or adjusting the savings for time-of-use (TOU) pricing or demand charges might be necessary.

Overall, the concept of a savings load shape highlights the importance of considering the timing and pattern of energy use when valuing the impacts of energy management activities.

4.20.3 Duck curves and negative pricing

The concept of duck curves and negative pricing can affect energy management in several ways. The duck curve refers to the shape of the electricity demand curve in regions with high penetration of solar energy. During the day, solar energy production is high, leading to a surplus of electricity that can drive down the market price. Then, as the sun sets and demand increases, electricity generation from non-solar sources must ramp up quickly to meet demand, resulting in a steep increase in the demand curve.

This phenomenon can create challenges for energy management, as it requires energy systems to be flexible and responsive to rapid changes in supply and demand. In particular, it can be difficult to manage energy storage systems, such as batteries, to ensure that they are charged during times of excess supply and discharged during times of peak demand. This is important because if energy storage is not properly managed, it can lead to excess generation during low-demand periods, resulting in negative pricing and wasted energy.

To address these challenges, energy managers may need to incorporate advanced technologies and strategies, such as demand response, energy storage management systems, and dynamic pricing mechanisms. They may also need to work closely with utilities and energy providers to better understand the dynamics of the electricity market and how to optimize energy use in the face of these challenges.

But what about you, the M&V professional? How should your approach adapt in the face of negative energy prices?

- **Use interval data:** By using interval data, you can get a more accurate picture of energy usage over time, including during periods of high renewable generation and low demand.
- **Turn to new performance metrics:** Traditional performance metrics such as kWh "savings" may not be appropriate in a scenario where negative prices occur. You can turn to new metrics such as peak load reduction, demand response participation, and carbon reduction to better reflect the value of energy management measures.
- **Use simulation models:** Simulation models can be used to predict the impact of renewable energy on the grid and identify potential issues with negative pricing. These models can help inform energy management strategies and provide a better understanding of the potential risks and benefits of different approaches.

The National Renewable Energy Laboratory[56] (NREL) provides various resources on renewable energy integration, including publications, webinars, and software tools, and the Department of Energy's Grid Modernization Initiative[57] shares information on grid modernization and the challenges of integrating renewable energy. The Association of Energy Engineers (AEE) also offers guidance and training.

[56]https://www.nrel.gov/grid/index.html
[57]https://www.energy.gov/gmi/grid-modernization-initiative

4.20.4 Fixed-rate fees and tariffs

As mentioned many times in this book, the way energy is generated, delivered, and priced is undergoing a vast transition. Assigned value to energy management requires consideration for the price paid for the energy. But in some jurisdictions, the total cost to deliver energy is taking on greater importance. New rate structures are being considered that contain a high fixed charge for access to the energy grid, coupled with variable rates for time of use.

A high fixed charge and lower electric rates may impact energy efficiency in several ways.

Firstly, a high fixed charge means that customers will be paying a larger portion of their electricity bill as a fixed fee, regardless of the amount of energy they consume. This can reduce the financial incentive for customers to invest in energy efficiency measures since they will still be paying a high fixed charge regardless of how much energy they use. This can make it harder for energy efficiency measures to be cost-effective and reduce the value of energy efficiency.

Secondly, lower electric rates can reduce the financial benefits of energy efficiency measures since the cost of electricity is already relatively low. In this case, even if customers are able to reduce their energy consumption through energy efficiency measures, the financial benefits may not be as significant since the savings on the energy charge may be relatively small.

Overall, the impact of a high fixed charge and lower electric rates on energy efficiency will depend on the specific details of the utility rates and the energy efficiency measures being considered. However, in general, a high fixed charge and lower electric rates can reduce the financial incentive for customers to invest in energy efficiency measures, which may make it harder to promote and achieve energy efficiency goals.

4.21 The M&V Plan and Report

The M&V plan delineates the steps and metrics for comparing actual energy consumption with hypothetical energy usage (counterfactual). It includes details on data collection, analysis, and reporting of results to relevant stakeholders. Adherence to a well-structured M&V plan ensures a systematic approach, guaranteeing credible and defensible outcomes. A comprehensive M&V plan should address the following points:

- **Goals:** Clearly articulate the objectives, including the intended energy impact, performance metrics, and verification procedures.

- **Baseline:** Specify the timeframe and methodology for determining baseline energy consumption.
- **Measurement methods:** Describe the measurement techniques, tools, and performance metrics that will be used to quantify the impact of energy management activities, including the frequency and duration of these measurements.
- **Data collection:** Detail the procedures for data collection, including responsibilities, sources, quality control, and data management. Discuss data archiving as well, specifying the location, backup strategy, and retention period. This is vital for future references and compliance with regulatory mandates.
- **Reporting requirements:** List the methods and requirements for conveying results to stakeholders, including the frequency and format of reports.
- **Verification procedures:** Define the methods and criteria for verifying energy management activities' measured impacts and performance metrics.
- **Contingency plans:** Include strategies for addressing unforeseen events or changes in facilities or equipment that might affect the results of the M&V.
- **Roles and responsibilities:** Determine the roles and responsibilities of all stakeholders, from the project team, to facility owners and staff, to the M&V professionals.
- **Budget:** Provide a budget for the M&V activities.
- **Schedule:** Furnish a detailed timeline with milestones, deliverables, and deadlines for the M&V program.
- **Reporting plan:** Outline a reporting plan that specifies the recipients, frequency, and content of the reports.

4.22 Reporting in Measurement and Verification (M&V) Projects

M&V reports convey the outcomes of energy efficiency projects to stakeholders. While the goals and contents of M&V reports can differ based on the project and stakeholders, they typically include the following components:

- **Verification of energy impacts:** Include a summary of the actual vs. anticipated impacts of energy management activities, accompanied by an explanation of the M&V methodology and assumptions.

- **Project performance assessment:** Evaluate how the energy efficiency project fared against its initial goals and objectives, analyzing any challenges faced and the solutions implemented to overcome them.
- **Cost–benefit analysis:** Incorporate a cost–benefit study that contrasts the expenses of the energy efficiency project with the energy impacts achieved. This should also address non-monetary benefits, such as enhanced comfort or productivity.
- **Conclusions and future directions:** Summarize the principal findings and conclusions and discuss recommended next steps, like additional M&V activities.

4.22.1 Relationship between M&V plans and reports

- **Strategy vs. documentation:** M&V plans lay out the overarching strategy and methodology for M&V, whereas M&V reports offer in-depth documentation of the actual impacts achieved.
- **Framework vs. presentation:** M&V plans provide a framework for data collection and analysis, while M&V reports present the analysis outcomes.
- **Metrics and baseline vs. comparison:** M&V plans set the performance metrics and baseline, whereas M&V reports contrast actual performance against these metrics and report the findings.
- **Roles and responsibilities vs. record of activities:** M&V plans define the roles and responsibilities of all participants, whereas M&V reports document the activities and contributions of these participants.
- **Blueprint vs. record of success:** M&V plans act as a blueprint for the entire project, while M&V reports describe its outcomes.
- **Timing:** M&V plans are generally formulated at the project's outset, while M&V reports are produced periodically throughout the project and upon its conclusion.

Both M&V plans and reports are necessary for a successful project, ensuring accountability, transparency, and accuracy.

4.22.2 Special considerations for reporting in utility program contexts

Reporting within the utility program framework involves special considerations with respect to program objectives, requirements, and reporting formats.

- **Alignment with program goals and requirements:** Ensure the M&V report aligns with the program's goals and requirements, such as reporting timelines, formats, and data quality standards.
- **Consistent reporting formats:** Utility programs frequently require the use of specific reporting formats for simplified data analysis and cross-project comparison.
- **Stakeholder engagement:** Utility programs often involve diverse stakeholders, such as program administrators, project developers, energy service companies, and customers. M&V reporting in this context may therefore have different audiences than in ESPC.
- **Continuous improvement:** Utility programs strive for continuous improvement and customer satisfaction. The M&V report should provide actionable insights and recommendations for program and project improvement grounded in the M&V analysis and findings.

4.22.3 Contractual alignment

M&V plans must align with the contracts they are part of. For this reason, you should develop the M&V plan concurrently with the project contract to ensure it truly represents the project's aims and objectives. The M&V plan's language should mirror the contract provisions, especially concerning methodology, data collection, and reporting. A lack of alignment with contract specifications means the M&V process will not be consistent with contractual obligations.

4.22.4 Utility programs

Utility programs usually impose unique requirements that influence M&V plans. For instance, utilities may set specific energy efficiency targets, data collection processes, and methodological and reporting guidelines that M&V plans must adhere to. Utility programs often include customer engagement and outreach initiatives to boost participation. As such, M&V plans must account for these customer engagement activities and ensure that calculations incorporate customer behavior or participation changes.

4.22.5 Disputes happen: avoiding, managing, and resolving them

A System for Avoiding Disputes

Despite our best efforts, disputes happen, whether they are employment-related disputes, regulatory or legal disputes, or contractual disputes, which may emerge before or after a contract is awarded.

That said, you *can* minimize the likelihood of disputes. In fact, avoiding disputes is a critical aspect of effective M&V project management.

Understanding the root causes of disputes is key to prevention. The most common are poor communication and perceived or actual deficiencies in performance. For example, if a contractor does not complete their scope of work or a client fails to meet their responsibilities, a dispute may arise. Similarly, equipment or system failures can lead to issues, whether it is the contractor-supplied equipment or the client's. Disputes can also emerge over costs and payments: the contractor might want to be paid more, or the client might want to pay less or delay payment. Inadequate O&M, changes made without approval, or inconsistencies in documentation of energy impacts can be fertile grounds for dispute as well.

But disputes often go beyond immediate communication failures and/or poor performance. A poorly written contract, for instance – with ambiguous specifications, unclear role definitions, inappropriate risk allocation, or vague dispute resolution clauses – can be a precipitating factor. Changes in staffing or management, inadequate or untrained staff, lack of documentation, and challenges introduced by new or complex technologies can further complicate the scenario.

To avoid disputes, you should watch for warning signs, like sudden changes in communication patterns, unforeseen site conditions, excessive requests for changes in work, payment delays, and unaddressed mistakes.

You should also optimize communication by designating clear roles and responsibilities for all parties, defining measurement procedures, and establishing effective systems for communication and documentation. A good M&V plan will concisely outline all these elements and other procedural aspects, such as baseline and performance metrics, data collection methods, and reporting requirements.

You can also employ other strategies to minimize the risk of disputes and maximize the project's chances of success. Namely:

- Contract stipulations and roles should be clear and unambiguous, ethical standards must be upheld, and any mistakes swiftly acknowledged and addressed.
- Regular meetings and progress reports should be scheduled to discuss any issues or concerns and ensure the project is progressing as planned.
- All parties involved should have access to the same data and reports, and there should be a transparent process for reviewing and verifying data.

- If not already stipulated in the contract, the M&V plan should define a dispute resolution process that can efficiently resolve any disagreements that may arise. This process should be communicated to all stakeholders.

If dispute prevention fails, four main methods can be employed to resolve disputes: negotiation, mediation, arbitration, and litigation.

- Negotiation – the submission and consideration of offers until an acceptable offer is agreed upon – is often the fastest and least expensive method, but it lacks a guaranteed resolution.
- Mediation brings in a neutral third party who helps the contending parties settle their dispute. This method encourages compromise and creative solutions, rather than forcing a win/lose outcome.
- Arbitration is the referral of a dispute to an impartial third party chosen by the disputants. They agree in advance to abide by the arbitrator's decision following a hearing. While arbitration can be less formal and faster than litigation, remember that its outcomes are binding and final.
- Lastly, litigation is a more formal and public process, usually more expensive but necessary when factual discovery is required or legal issues may dictate the outcome.

As an additional measure, dispute resolution boards, composed of impartial professionals with relevant expertise, can be employed from the start of a major project. They offer recommendations for avoiding disputes and act as a standing panel to address disputes when they arise.

Examples of Disputes and Their Resolutions

Dispute over Baseline Energy Consumption

Disputes over baseline energy consumption typically arise due to disagreements between the facility owner and the energy service provider regarding the selection of the baseline model, choice of variables, or accuracy of data used.

To resolve such disputes, the first step is to review the contractual agreement to determine if there are any provisions for resolving disputes related to the baseline model. If the contractual agreement does not provide a clear path to resolution, conducting an independent review of the baseline model by a third-party expert may be necessary. In cases where an agreement cannot be reached through the above means, engaging a neutral third party in mediation or arbitration is a viable option. If all else fails, litigation may be the final recourse.

Non-routine Adjustment Dispute

In the case of a non-routine adjustment dispute, which involves disagreements over whether a change in the facility or system's operation should be considered a non-routine adjustment, refer to the M&V plan for definitions and procedures regarding non-routine adjustments. If a dispute arises, an independent third party may need to be engaged to review the M&V plan and data.

Dispute in Valuation of Avoided Energy

Disputes in the valuation of avoided energy stem from differing opinions about which methods and assumptions should be used. To avoid these disputes, have clear and agreed-upon procedures for calculating the value of avoided energy. This includes documenting the assumptions and inputs used in the calculation in the M&V plan. If a dispute arises despite this, an independent third party can be engaged to review the calculations. Alternatively, parties may choose to negotiate and try to reach a compromise.

Demand Reduction Disputes

For demand reduction disputes, which involve disagreements regarding the calculation of demand reduction, a thorough review of the M&V plan and other documentation related to the project is required. This includes data on the baseline and post-retrofit periods, calculation methodologies used, and any assumptions made. If the dispute cannot be resolved internally, independent third-party experts could be brought in to provide an objective assessment. Once all parties have reviewed the data and any third-party reports, negotiations can begin to resolve the dispute. This may entail revising the M&V plan, adjusting performance metrics, or renegotiating the contract. If negotiations fail, dispute resolution mechanisms such as mediation or arbitration can be used.

* * *

Disputes in M&V processes can be minimized or managed effectively by establishing clear roles, defining measurement protocols, setting up effective communication channels, and implementing a well-defined dispute resolution process – so when disputes *do* arise, there is a structured approach to resolution. As expected, collaboration and open communication are the linchpins of successful dispute avoidance, management, and resolution. And throughout it all, maintaining humanity – listening, empathy, dignity, and forgiveness – is of the utmost importance.

4.23 Third-party M&V

The proper role of third-party M&V is to provide an independent, unbiased assessment of the impact of energy management activities implemented through an energy-savings performance contract or energy efficiency project. This helps ensure that the reported impacts are accurate, reliable, and verifiable. The third-party M&V provider should have the expertise and experience necessary to design and implement a comprehensive M&V plan, including data collection, analysis, and reporting. They should also have a thorough understanding of industry standards and best practices. They should work closely with the ESCO and the client to ensure that the M&V plan is tailored to the project's specific needs and that all parties find it satisfactory.

Third-party M&V for contracts is typically more comprehensive and customized than M&V for utility programs. In utility programs, the M&V process is often standardized and designed to cover a wide range of projects, focusing on verifying energy impacts for incentive payments. Third-party M&V for contracts, on the other hand, involves developing a customized plan specific to the project, which may include site-specific metering and data collection, calibrated simulation models, and comprehensive analysis of uncertainties and risks associated with the project.

A facility owner and an ESCO may not want to do the M&V themselves for several reasons:

- They may lack the expertise or resources to perform the M&V work effectively.
- Performing the M&V themselves may raise questions about the impartiality and accuracy of the results.
- Third-party M&V can provide a valuable external perspective and identify areas for improvement that may not be apparent to the facility owner or ESCO.

That said, third-party M&V may not be useful or feasible if any of the following conditions apply:

- **Small projects:** For smaller projects, the cost of third-party M&V may not be justified, and the facility owner may prefer to rely on in-house staff to perform the M&V.
- **Limited budget:** If there is a limited budget for a project, third-party M&V may not be feasible, and the facility owner may opt for a simpler M&V plan that can be managed in-house.

- **Low risk:** In some cases, the risks associated with a project may be low, and the facility owner may decide that the added cost of third-party M&V is not warranted.
- **Owner experience:** If the facility owner has significant experience in M&V and is confident in their ability to perform the work, third-party M&V may be unnecessary.
- **Regulatory exemptions:** Some regulatory bodies may exempt certain projects from third-party M&V requirements based on size, scope, or other factors.

It is important to assess the unique circumstances of each project to determine whether third-party M&V is necessary and cost-effective.

4.24 The Future

The current set of methods for M&V has been in use for the past three decades with much success. Indeed, in many jurisdictions, these "classic" methods will continue to be viable for decades to come. However, the world of energy and the definition of energy management are evolving. The core concept of a stable counterfactual baseline may not apply in many current and future situations. In order to continue to support energy management with tools to cost-effectively quantify its value, the M&V community will need to continue to evolve as well.

4.24.1 The energy transition[58]

As we have discussed, M&V's standard guidance documents arose in an era where the assumption of a stable baseline (from which to implement the counterfactual method) was reasonable. But is that really now the case? One could plausibly state that for many facilities, the baseline energy use pattern will not reflect future use.

For this, you can blame the "energy transition."[59] Repeating the theme of being cautious about terminology, the "energy transition" covers many issues pertinent to energy management. Renewable energy and energy storage (behind the meter and grid level), electric vehicles, real-time electricity pricing, fuel switching (electrification), and carbon reduction all contribute to a fundamental change in how energy is generated, distributed, and used.

[58]https://www.mckinsey.com/search?q=energy%20sustainability%20transition

[59]https://www.spglobal.com/en/research-insights/articles/what-is-energy-transition#:~:text=Energy%20transition%20refers%20to%20the,well%20as%20lithium%2Dion%20batteries.

As the energy transition proceeds, the technologies and policies that shape our energy systems are evolving rapidly. This promises many opportunities, but it also poses new challenges to the M&V professional:

- **Changing energy mix:** As more renewable energy sources are integrated into the grid, the variability and intermittency of these sources can make it difficult to establish accurate baselines for the impacts of energy management activities. M&V professionals must develop new methods to accurately account for the impact of renewable energy sources on energy usage.
- **Rapid technological change:** The pace of technological change in the energy sector can make it difficult for M&V professionals to keep up. New technologies, such as energy storage systems and smart grid technologies, may require new M&V protocols and tools to accurately measure their impact on energy consumption.
- **Increasing complexity:** As energy systems become more complex, M&V professionals may face challenges in accurately measuring energy usage across multiple systems and components. They may need to work with other professionals to develop more sophisticated methods for M&V.
- **Data quality and availability:** M&V relies heavily on accurate and timely data. However, data quality and availability can be challenging, particularly in older buildings or regions with limited data infrastructure access. M&V professionals must work with facility owners, utilities, and other stakeholders to ensure data is collected, stored, and analyzed to support effective M&V.

Modern M&V, as defined in "the big three" – FEMP M&V, IPMVP, and ASHRAE Guideline 14 – M&V guidance documents, is now in its fourth decade. M&V has proven effective for many of the current needs of ESPC and utility and government programs, albeit with many issues still unresolved. An M&V professional has options for practicing the fundamentals of M&V anywhere in the world. And while the M&V professional will almost certainly not be an expert in every aspect of M&V, the options for lifelong learning are greater than ever.

These are other technological advancements that have the potential to impact the field of M&V:

- **Internet of Things (IoT):** IoT refers to the interconnection of various devices and sensors through the Internet. This technology can help M&V

professionals collect more accurate and timely data, which can be used to improve energy performance and optimize operations.

- **Machine learning:** These technologies enable M&V professionals to analyze and make sense of large amounts of data. With the help of data science and machine learning techniques, M&V professionals can develop more accurate models and predictions for energy performance.
- **AI:** AI technologies can help automate M&V processes and improve accuracy. AI can assist in real-time monitoring and adjusting of energy usage and identify data patterns that humans might overlook.
- **Digital twins:** A digital twin is a digital replica of a physical asset. By creating a digital twin of an energy system, M&V professionals can simulate different scenarios and test the impact of changes to the system before implementing them in the physical world.
- **Blockchain:** Blockchain is a distributed ledger technology that can securely and transparently track energy usage and transactions. This technology can potentially improve the accuracy and transparency of M&V data.
- **Smart contracts:** Smart contracts are self-executing contracts with the terms of the agreement between buyer and seller being directly written into lines of code.
- **Bayesian calibration:** Bayesian calibration is a statistical technique used to estimate uncertainty in model parameters.

The evolving energy and technology transitions will require M&V professionals to adapt. The old world of stable baselines will be harder to find. A new world full of Digital twins, EVs, PV, DER, DR, smart meters, flex markets, etc., will require new tools and continuing education.

4.24.2 Flex markets

As more renewable energy sources are integrated into the grid, it becomes increasingly challenging to accurately measure and verify the impact of energy management activities, as the baseline against which they are measured is constantly shifting. Flex markets manage the variability and intermittency of renewable energy sources by enabling grid operators and energy market participants to buy and sell flexible capacity and demand responses in real time. These flex markets require real-time monitoring and verification of energy consumption and generation – but thankfully, the energy management industry is becoming more sophisticated in its use of sensors, meters, and analytics to provide accurate and reliable data. The industry is also incorporating new machine learning and artificial intelligence techniques to allow for more efficient and effective analysis of large datasets.

For M&V, one of the key challenges in serving flex markets is the need to accurately measure and verify the impact of distributed energy resources (DERs) on the grid. M&V is evolving to address this challenge by developing new methods for measuring and verifying the performance of DERs, including solar panels, wind turbines, and energy storage systems. This includes the development of new performance metrics, such as ramp rates and response times, that can be used to measure the ability of DERs to respond to grid events and maintain grid reliability.

Clearly, the development of flex markets represents an exciting opportunity for the M&V community to innovate and adapt to the changing energy landscape and play a key role in transitioning to a more flexible, resilient, and sustainable energy system.

4.24.3 Integrated policy framework

The policies around grid integration and climate are changing how we use and value energy – and the value of managing it.

While we currently lack an integrated policy framework that would bring together many of the initiatives that impact our energy systems, an M&V professional should be aware of opportunities to bundle value from various programs such as energy efficiency, demand reduction (DR), distributed energy resource (DER), and carbon.

While some of these changes will not require entirely new M&V methods, some will. The concept of a stable baseline that can be used to compare to actual future use could become inoperative in many areas in the near future – if it is not already.

This will require that the M&V professional stay with global trends and in their specific community of practice. All of this makes the role of the M&V professional challenging and exciting in the years ahead.

4.24.4 The future of M&V

Finally, while speculative, one can imagine a future world where "performance" becomes much more important than "savings." Performance targets are often more straightforward and easier to communicate to stakeholders, as they involve setting a specific goal to be achieved, such as a certain level of emission reductions or reduced energy consumption, by a certain date.

Compliance with the target can be tracked through simple metrics, such as energy bills or emissions data, which tends to be readily available.

Performance targets like net zero emissions – a state in which the amount of greenhouse gas emissions produced is equal to the amount removed from the atmosphere, either through natural processes or technological solutions – are becoming more popular as a way to drive meaningful action on climate change. They provide a clear and measurable goal for governments, businesses, and individuals to work toward and can help to drive innovation and investment in low-carbon technologies and practices. And they do not necessarily involve using M&V to measure and verify the actual impacts of energy management (or emission reduction) activities.

If net zero energy and net zero carbon policies proliferate, future energy systems (buildings, anyway) may be required to meet short-term or annual targets. However, M&V could still play a role in verifying that a building has achieved its net zero energy or carbon target. For example, an M&V professional could monitor energy use in the building and verify that it is operating at net zero energy or carbon levels. Similarly, M&V could be used to verify the performance of specific energy efficiency or renewable energy measures implemented in the building to achieve the net zero target.

Whatever unfolds in the coming years, the M&V professional must adapt to the new realities. The old methods may need to be abandoned. New methods of analysis will require new skills. But there will always be a role for the M&V professional who can navigate the changing landscape and successfully settle projects.

4.25 Additional Training and Information

Advancing and implementing future energy and climate policies will require a range of skills and expertise, both technical and non-technical. Some of the key areas where additional training may be needed include:

- **Technical skills:** Energy and climate policies are complex and require a strong technical understanding of energy systems, carbon accounting, and data analysis. This includes expertise in fields such as engineering, economics, and environmental science, as well as proficiency in the use of specialized software and tools for energy modeling and analysis.
- **Policy analysis and development:** Energy and climate policies are developed and implemented through a complex policymaking process

that involves stakeholders from various sectors and levels of government. As such, training in policy analysis and development, including skills in stakeholder engagement, policy design, and policy evaluation, will be critical.

- **Communication and advocacy:** Effective communication and advocacy are essential for promoting the adoption of energy and climate policies and engaging stakeholders. Training in communication and advocacy can help individuals develop skills in messaging, media relations, and public speaking, among other areas.
- **Leadership and collaboration:** Advancing and implementing energy and climate policies will require strong leadership skills, as well as the ability to work collaboratively with stakeholders from diverse backgrounds. Training in leadership and collaboration can help individuals develop skills in team-building, conflict resolution, and strategic planning.
- **Cultural competency:** Energy and climate policies are global issues that affect people from diverse cultural backgrounds. As such, training in cultural competency, including an understanding of cultural norms, values, and beliefs, will be essential for effective stakeholder engagement and policy implementation.

Training and professional development opportunities will be critical to ensure that you have the knowledge and skills needed to drive progress toward a sustainable future.

4.25.1 A global guild

The growth of an international community of M&V practitioners, a "global guild," is undeniable, with established international organizations such as AEE and EVO already playing a key role. Their active involvement in M&V, an increasingly important component of energy efficiency and climate policies, has laid a firm foundation for the future of the field.

Both academia and industry have valuable roles to play in the future of M&V. Academia conducts research on M&V methods and techniques and provides training and education for M&V practitioners.

Industry provides practical insights and experience from real-world projects. Industry practitioners can share their experiences and best practices, identify emerging trends and challenges, and work collaboratively with academia to develop new protocols and standards.

Beyond the development of standards, guidance, and protocols, these and other organizations are actively involved in facilitating peer-to-peer learning and support. They have established platforms for knowledge exchange and collaboration, enabling professionals from different regions and sectors to learn from one another's experiences and support each other on challenging projects.

They also promote the significance of M&V in energy efficiency and climate policies, actively raising awareness among potential stakeholders like policymakers and facility owners of the benefits of accurate and reliable M&V.

Professional development and recognition are other areas where these organizations are making an impact. They offer training and certification programs, thus solidifying M&V as a recognized profession. They also help practitioners gain greater recognition and credibility within their organizations and the broader community.

In sum, the global M&V landscape is rich with resources and organizations ready to support practitioners, no matter where they are located. These international bodies have ensured that M&V expertise is not bound by geography. As an M&V professional, you should tap into these resources, learn from others' knowledge and experiences, and actively participate in this global guild.

5

Conclusion

The term "measurement and verification" has been in wide use since around 1990. It is currently widely defined as the set of concepts that are used to account for the impact of energy management projects and programs.[2]

M&V methods, derived from science, engineering, statistics, and econometrics, are described in various "standard" documents. Each of these documents takes a slightly different approach to defining M&V for a particular purpose. Each comes from a unique perspective.

In this book, I have attempted to demystify some of these perspectives and terminology, allowing the M&V professional to have access to the widest possible set of tools and methods. The main goal is to allow anyone using M&V to better understand and hence implement successful, cost-effective M&V. But I am also hoping that the M&V field will be open to contributions from a wide range of experts who are familiar with the general challenges faced when implementing the counterfactual method.

There are many potentially beneficial uses of M&V, but the primary purpose is to establish methods whereby stakeholders can agree on the calculated results. There is no "negawatt" meter. While fun to discuss, this concept is nevertheless misleading and somewhat disrespectful to the professionals who must use all their skills to guide projects through to successful settlement.

M&V is not a panacea for the problem of quantifying the absence of energy use. There is no single tool that can be deployed to *unequivocally* report a quantity of energy that is not consumed. M&V is a set of concepts that package logical assumptions, based on measured data and repeatable analysis to obtain an estimated result. There are methods, and the results generated from those methods. Experts can, and will, disagree.

The most important concept in M&V is that reasonable people can work cooperatively to establish and conduct M&V in a way that leads to successful settlement. That is, reported results that are acceptable to all parties.

The role of the M&V professional is to foster cooperation in the interest of attaining an acceptable result at an acceptable cost. That will likely mean assembling a team of technical specialists. But it will also mean being able to communicate with lawyers, bankers, regulators, and others to establish expectations early and ensure that there are no surprises. You will need to understand the process and anticipate roadblocks and challenges. In this way, you can proactively plan for these, rather than responding under less agreeable circumstances.

Throughout the M&V process, there will be many decisions requiring judgment. Professional judgments abound in many areas of life and are the foundation of every true profession. This is not a flaw. It is a reality. Embrace the creativity that is allowed and required. It is part of the fun.

A simple list of M&V activities might be:

- To understand the possibilities and limitations of the general M&V toolkit
- To work cooperatively with all stakeholders to agree on the best method to deploy for a given situation
- To create a cost-effective M&V plan
- To oversee the proper implementation of the plan
- Determine the baseline and establish the baseline model
- Verify the proper installation of the measures (capability to perform)
- Implement M&V activities in the reporting period, including identifying non-routine events and performing non-routine adjustments, as necessary
- Manage communication throughout the process
- Report the results, including all the relevant data and process steps

As succinct as those steps are, the range of skills needed to perform them is vast. An M&V professional will frequently be utilizing advanced technical and analytical resources. There is a limitless amount of learning to stay current with all of the existing energy technologies and analysis tools. And they are evolving.

The energy infrastructure of our planet is undergoing rapid transformation. The concept of "savings" is being complicated by the profusion of new sources of energy and energy demands. From electrification to energy storage to flexibility, the need for accountability is greater than ever – and more difficult. "Standard" M&V based on a "fair and relevant" (counterfactual) baseline will be around for a while. But much more sophisticated analysis is already needed in many jurisdictions. Fortunately, there is a growing community of experts who can perform advanced methods utilizing artificial intelligence, machine learning, digital twins, and Bayesian analysis. The Internet of Things and artificial energy intelligence could revolutionize M&V, if we learn how to harness them.

To fully address all these challenges, we must expand our professional network to include the many thousands of potential partners in our pursuits. Fortunately, the Internet has given us many tools to develop new communities of practice. Take advantage of them.

Working with a wide range of experts, learning new skills, and enjoying the satisfaction of completing successful projects are just some of the rewards of becoming an M&V professional. You can also take credit for being an important part of the solution to the global challenge of bringing clean, sustainable energy to all.

Good luck with your projects!

Appendix I: Evolution of IPMVP Option Definitions

Year	Option	Description	How savings are calculated
1996 (NEMVP)	A	Verifying ECM has potential to perform and generate savings	Metering: none or short-term periodic.
	B	Verifying ECM has potential to perform; verifying actual performance by end use	Continuous in post-installation at system level.
1997 (IPMVP)	A	Focuses on physical assessment of equipment changes to ensure the installation is to specification. Key performance factors (e.g., lighting wattage or chiller efficiency) are determined with spot or short-term measurements and operational factors (e.g., lighting operating hours or cooling ton-hours) are stipulated based on analysis of historical data or spot/short-term measurements. Performance factors and proper operation are measured or checked annually.	Engineering calculations using spot or short-term measurements, computer simulations, and/or historical data.
	B	Savings are determined after project completion by short-term or continuous measurements taken throughout the term of the contract at the device or system level. Both performance and operations factors are monitored.	Engineering calculations using metered data.
2002 (IPMVP)	A Partially measured retrofit isolation	Savings are determined by partial field measurement of the energy use of the system(s) to which an ECM was applied, separate from the energy use of the rest of the facility. Measurements may be either short-term or continuous. Partial measurement means that some but not all parameter(s) may be stipulated, if the total impact of possible stipulation error(s) is not significant to the resultant savings. Careful review of ECM design and installation will ensure that stipulated values fairly represent the probable actual value. Stipulations should be shown in the M&V plan along with analysis of the significance of the error they may introduce.	Engineering calculations using short-term or continuous post-retrofit measurements and stipulations.

Year	Option	Description	How savings are calculated
	B Retrofit isolation	Savings are determined by field measurement of the energy use of the systems to which the ECM was applied, separate from the energy use of the rest of the facility. Short-term or continuous measurements are taken throughout the post-retrofit period.	Engineering calculations using short-term or continuous measurements.
2007 (IPMVP)	A Retrofit isolation: key parameter measure-ment	Savings are determined by field measurement of the key performance parameter(s) that define the energy use of the ECM's affected system(s) and/or the success of the project. Measurement frequency ranges from short-term to continuous, depending on the expected variations in the measured parameter, and the length of the reporting period. Parameters not selected for field measurement are estimated. Estimates can be based on historical data, manufacturer's specifications, or engineering judgment. Documentation of the source or justification of the estimated parameter is required. The plausible savings error arising from estimation rather than measurement is evaluated.	Engineering calculation of baseline and reporting period energy from: short-term or continuous measurements of key operating parameter(s); estimated values. Routine and non-routine adjustments as required.
	B Retrofit isolation: all parameter measure-ment	Savings are determined by field measurement of the energy use of the ECM-affected system. Measurement frequency ranges from short-term to continuous, depending on the expected variations in the savings and the length of the reporting period.	Short-term or continuous measurements of baseline and reporting period energy, and/or engineering computations using measurements of proxies of energy use. Routine and non-routine adjustments as required.

Year	Option	Description	How savings are calculated
2010 (IPMVP)	A Retrofit isolation: key parameter measure-ment	Savings are determined by field measurement of the key performance parameter(s) that define the energy use of the ECM's affected system(s) and/or the success of the project. Measurement frequency ranges from short-term to continuous, depending on the expected variations in the measured parameter and the length of the reporting period. Parameters not selected for field measurement are estimated. Estimates can be based on historical data, manufacturer's specifications, or engineering judgment. Documentation of the source or justification of the estimated parameter is required. The plausible savings error arising from estimation rather than measurement is evaluated.	Engineering calculation of baseline and reporting period energy from: short-term or continuous measurements of key operating parameter(s); estimated values. Routine and non-routine adjustments as required.
	B Retrofit isolation: all parameter measure-ment	Savings are determined by field measurement of the energy use of the ECM-affected system. Measurement frequency ranges from short-term to continuous, depending on the expected variations in the savings and the length of the reporting period.	Short-term or continuous measurements of baseline and reporting period energy, and/or engineering computations using measurements of proxies of energy use. Routine and non-routine adjustments as required.
2012 (IPMVP)	A Retrofit isolation: key parameter measure-ment	Savings are determined by field measurement of the key performance parameter(s) that define the energy use of the ECM's affected system(s) and/or the success of the project. Measurement frequency ranges from short-term to continuous, depending on the expected variations in the measured parameter and the length of the reporting period. Parameters not selected for field measurement are estimated. Estimates can be based on historical data, manufacturer's specifications, or engineering judgment. Documentation of the source or justification of the estimated parameter is required. The plausible savings error arising from estimation rather than measurement is evaluated.	Engineering calculation of baseline and reporting period energy from: short-term or continuous measurements of key operating parameter(s); estimated values. Routine and non-routine adjustments as required.

Year	Option	Description	How savings are calculated
	B Retrofit isolation: all parameter measurement	Savings are determined by field measurement of the energy use of the ECM-affected system. Measurement frequency ranges from short-term to continuous, depending on the expected variations in the savings and the length of the reporting period.	Short-term or continuous measurements of baseline and reporting-period energy, and/or engineering computations using measurements of proxies of energy use. Routine and non-routine adjustments as required.
2014 (Core concepts)	A Retrofit isolation: key parameter measurement	Savings are determined by field measurement of the key parameter(s), which define the energy use of the ECM's affected system(s) or the success of the project. Measurement frequency ranges from short-term to continuous, depending on the expected variations in the measured parameter and the length of the reporting period. Parameters not selected for field measurements are estimated. Estimates can be based on historical data, manufacturer specifications, or engineering judgment. Documentation of the source or justification of the estimated parameter is required. The plausible saving error arising from estimation rather than measurement is evaluated.	Engineering calculation of baseline period energy and reporting period energy from: short-term or continuous measurements of key operating parameter(s); estimated values. Routine and non-routine adjustments as required.
	B Retrofit isolation: all parameter measurement	Savings are determined by field measurement of the energy use of the ECM affected system. Measurement frequency ranges from short-term to continuous, depending on the expected variations in savings and length of the reporting period.	Short-term or continuous measurements of baseline and reporting period energy, or engineering computations using measurements of proxies of energy uses. Routine and non-routine adjustments as required.

Year	Option	Description	How savings are calculated
2016 (Core concepts)	A Retrofit isolation: key parameter measure-ment	Savings are determined by field measurement of the key parameter(s) that define the energy consumption and demand of the ECM's affected system(s) or the success of the project. Measurement frequency ranges from short-term to continuous, depending on the expected variations in the measured parameter and the length of the reporting period. Parameters not selected for field measurements are estimated values. Estimates can be based on historical data, manufacturer specifications, or engineering judgment. Documentation of the source or justification of the estimated value is required. The plausible saving error arising from estimation rather than measurement is evaluated.	Engineering calculation of baseline period energy and reporting period energy from: short-term or continuous measurements of key parameter(s); estimated values. Routine and non-routine adjustments as required. Key parameter(s) measured during both baseline and reporting period.
	B Retrofit isolation: all parameter measure-ment	Savings are determined by field measurement of the energy consumption and demand and/or related independent or proxy variables of the ECM affected system. Measurement frequency ranges from short-term to continuous, depending on the expected variations in savings and length of the reporting period.	Short-term or continuous measurements of baseline and reporting period energy, or engineering computations using measurements of proxies of energy consumption and demand. Routines and non-routine adjustments as required.
2022 (Core concepts)	A Retrofit isolation: key parameter measure-ment	Savings are determined by field measurement of the key parameters(s) that define the energy consumption and/or demand of the EEM's affected system(s). Performance indicators may also be defined and measured to ensure the success of the project. Measurements range from periodic short-term to continuous long-term, depending on the expected variations in the key parameter(s). Parameters not selected for field measurements are estimated values. Estimates can be based on historical data, manufacturer specifications, or engineering judgment. Documentation of the source and justification of the estimated value is required. The plausible savings error arising from estimation rather than measurement is evaluated and acceptable to all stakeholders.	Calculation of baseline period energy and reporting period energy from periodic short-term energy measurements, or from periodic short-term or continuous measurements of key parameter(s) and from estimated values. Routine and non-routine adjustments as required. Key parameter(s) must be measured during both the baseline and reporting period.

Year	Option	Description	How savings are calculated
	B Retrofit isolation: all parameter measurement	Savings are determined by continuous field measurement of the energy consumption and/or demand or validated proxy variables and the related independent variables of the EEM-affected system. Measurements range from periodic short-term to continuous long-term, depending on the expected variations in the key parameters.	Determination of baseline period energy based on short-term or continuous measurements of baseline and reporting period energy, or on engineering computations using measurements of proven proxies of energy consumption or demand. Routine and non-routine adjustments as required.

References to River Pub E Library

Handbook of Energy Audits

https://elibrary.riverpublishers.com/ebook_details.php?id=864

Project Management

https://elibrary.riverpublishers.com/ebook_read.php?id=799

Energy Calculations

https://elibrary.riverpublishers.com/ebook_read.php?id=953

Energy Management Handbook 9th Edition

https://elibrary.riverpublishers.com/ebook_read.php?id=828

Uncertainty Examples (Measurement)

https://www.wasyresearch.com/measurement-uncertainty-estimations-gum-method/

Counterfactual and Causal Inference

https://edisciplinas.usp.br/pluginfile.php/3984640/mod_resource/content/2/%5BStephen_L._Morgan%2C_Christopher_Winship%5D_Counterfa%28BookFi.org%29%20%281%29%281%29.pdf

Glossary

Above code: Exceeding or more than what is required; this typically refers to energy impacts or efficiencies that exceed a baseline energy performance.

Accuracy (vs. precision): How close a reported value (measurement) is to the "true" value. Accuracy is relevant for many measurements that comprise M&V activities.

Additionality (energy efficiency programs): The energy management impacts caused by actions beyond what would have occurred in the absence of the energy efficiency program.

Adjustments: Modifications made to baseline energy models to account for changes in conditions such as weather, occupancy, and production levels.

Adjusted baseline model: A model developed using data from the baseline period and incorporating routine and non-routine adjustments, against which an energy conservation measure's actual performance is compared.

AEE: Association of Energy Engineers.

Algorithm (see also governing equations): A process or set of rules to be followed in calculations or other problem-solving operations, especially by a computer.

Apparent power: See "Power factor."

Approach (see also "Method"): Approaches and methods are two terms that are often used in the context of problem-solving, research, and learning. While the two terms are related, they are not interchangeable. An approach is a general way of thinking about or addressing a problem or situation. It is a broad perspective or mindset that guides how you approach a task or challenge. An approach can be thought of as a set of guiding principles that inform your actions and decisions.

ASHRAE: American Society of Heating, Refrigerating and Air-Conditioning Engineers.

Asset: Any equipment, device, system, or infrastructure component that consumes, generates, stores, or manages energy – for example, HVAC systems, solar panels, batteries, building management systems, and lighting fixtures.

Asset score: A numerical representation of the potential performance of an energy asset (or set of assets), determined in accordance with a standardized methodology and used for comparison.

Attribution (as in "risk attribution"): The process of identifying and assigning sources or factors that contribute to the variability or potential downside in a particular metric or outcome. In energy management, for example, risk attribution might involve determining which equipment or processes contribute most to energy usage variability or potential system failures.

Audit (predict compare diagram): A systematic assessment of energy consumption patterns within a facility or system to document existing conditions, identify inefficiencies, and recommend opportunities for energy management activities.

Advanced metering infrastructure (AMI): Utility metering infrastructure that provides interval data – that is, data collected at regular intervals, such as every 15 or 30 minutes – on energy consumption.

Avoided energy use: The estimated impact of an energy efficiency project under the conditions experienced in the reporting period, calculated by comparing the actual energy consumption after the project is implemented to the counterfactual (estimated energy consumption or cost that would have occurred in the absence of the project), accounting for changes in external factors that may affect energy consumption or cost, such as weather patterns or occupancy levels.

Avoided costs: The monetary "savings" resulting from avoided energy use.

Backcasting: A method of calculating energy impacts, which uses data from the reporting period to specify a model and predict the baseline period energy consumption.

Balance point (change point): The outdoor temperature at which a facility's heat loss (or gain) is equal to its internal heat gains (from occupants, lights, equipment, etc.). At this temperature, the heating or cooling system does not need to operate to maintain the desired indoor temperature.

Bayesian statistics: An approach that treats probability as a measure of belief or uncertainty, rather than as a frequency of occurrence. Bayesian models in M&V use a prior probability distribution, which represents existing knowledge or assumptions about the variables in the model, and update this distribution based on observed data to estimate the posterior probability distribution.

Baseline: A characterization of the energy use and operating conditions during a specified period prior to implementation of an energy project.

Baseline period: The period of time selected to document the baseline conditions and create a baseline model.

Benchmarking (general): The process of comparing the energy performance of a building or facility to a set of established standards or benchmarks.

Building performance standard: an outcome-based policy or law aimed at reducing the carbon impact of the built environment by requiring existing buildings to meet energy – or GHG emissions-based performance targets.

Building energy asset score: A standardized tool for calculating a building's energy performance based on physical characteristics such as building envelope, HVAC systems, and lighting. The score provides a snapshot of a building's energy efficiency and indicates potential areas for improvement.

Calibration (model): Adjusting the inputs of a model to match the actual energy consumption of a facility in the baseline period.

Calibration (instruments): It is the process of configuring an instrument to provide a result for a sample within an acceptable range.

Causal/causality: The relationship between cause and effect, where a change in one variable or event (the cause) produces a change in another variable or event (the effect).

CEM (Certified Energy Manager): An individual who optimizes the energy performance of a facility, building, or industrial plant. The CEM is a systems integrator for electrical, mechanical, process, and building infrastructure, analyzing the optimum solutions to reduce energy consumption in a cost-effective approach.

Change point: See "Balance point."

CMVP (Certified Measurement and Verification Professional): An individual who manages or performs internationally recognized methods of quantifying the energy usage impacts of energy management activities.

Codes: Building codes and energy codes refer to regulations and standards that set specific requirements for the energy efficiency of new buildings, renovations, equipment, and appliances, and sometimes the operation of buildings and structures.

Code baseline: The minimum efficiency standards that should have been in place at the time of implementation. Used in some utility programs, rather than an "existing conditions" baseline, to protect the interests of utility ratepayers; so they are not subsidizing projects where the facility owner should have already installed more efficient equipment.

Coefficient (see also "Parameter"): A numerical value that is multiplied with a variable or term in a mathematical expression. It represents the scaling or proportionality factor by which the variable or term is modified within the context of the equation or formula.

Commissioning: The process of verifying and documenting the proper performance of facilities, systems, and equipment.

Comparison group: In a quasi-experimental approach, the researcher compares the impact of an intervention on the experimental group to a comparison group that is similar to the experimental group in all respects except for the independent variable.

Confidence: The degree of certainty that can be placed in the accuracy or reliability of a particular measurement or estimate.

Confidence interval (see also "Credible interval): A range within which a parameter is likely to fall with a certain level of confidence. It gives an idea of the uncertainty around a measurement or estimate.

Conservation: Conservation involves the careful and responsible management of resources to prevent waste, depletion, or harm to the environment.

Contract: A legally binding document that specifies the terms and conditions of a project, including the scope of work, the expected impact of the energy management activities, the payment structure, and other important details.

Correlation coefficient: A statistical measure ranging from −1.0 to 1.0 that represents how positively or negatively correlated variables are.

Counterfactual analysis/method: The counterfactual method is a conceptual and analytical approach used in various fields, including philosophy, social sciences, and artificial intelligence. It involves reasoning about hypothetical scenarios that could have occurred if certain conditions were different from what actually transpired. In essence, it is a way of exploring what might have happened if specific events or conditions had taken an alternative course.

Credible interval: See "Confidence interval."

Current transformer (CT): A device used to measure and monitor the current flowing through an electrical conductor. It reduces high currents to a lower, safer level for measurement or protection equipment. CTs are commonly used in metering and protective relays in the electrical power industry. They provide an isolated low current output proportional to the high current in the main circuit (common types include solid core, split core, and coils).

Custom program: A utility program that gives customers incentives for implementing energy efficiency measures not covered by standard rebate programs. These measures are often tailored to the specific needs of the customer and can include things like advanced lighting controls, building automation systems, or customized HVAC solutions. The customer and utility work together to identify energy-saving opportunities and determine the appropriate incentive amount. The incentive is typically based on the estimated energy impact of the project.

CV(RMSE): Coefficient of variation of root mean squared error. In a regression analysis of energy data, CV(RMSE) is an indicator of the confidence in the model's predictive capabilities.

Data access[1].: Utilities have access to vast amounts of data on their customers' energy use and behaviors. Making this data accessible to customers can help them better manage energy use in homes, large buildings, and entire communities.

Data security: The process of safeguarding digital information throughout its entire lifecycle to protect it from corruption, theft, or unauthorized access.

Deemed savings (see "Stipulation" and "TRM"): Predetermined estimates of the avoided energy use achieved by energy conservation measures.

Denominator: The reference value or the base quantity against which another metric (the numerator) is being compared or normalized.

Dependent variable: A variable presumed to be influenced by an independent variable.

[1]https://database.aceee.org/state/data-access#:\sim:text=Utilities%20have%20access%20to%20vast,large%20buildings%2C%20and%20entire%20communities

Descriptive statistics: The process of analyzing data in a dataset in order to describe and offer meaningful information about the data, including possible relationships among data points.

Digital twin: A virtual representation of a physical system that can be used to simulate the behavior of the asset or system, predict its performance, and identify potential problems before they occur.

Dispute resolution: A process, usually stipulated in the contract or M&V plan, for resolving any disagreements that may arise.

Domains: It is possible to think of the work of the M&V professional as spanning four domains, or areas of practice: regulatory, contractual/programmatic, economic, and physical (engineering).

Dual fuel: Accounting for mixed fuel inputs (gas, electric, etc.).

Dual baseline: The dual baseline reflects the difference between the "savings" that should be credited for the initial years of installation based upon the pre-existing or replaced equipment vs. the "savings" credit in later years that should be based upon an eventual pre-existing equipment replacement assumed to occur if the measure had not been installed as part of the program.

Early retirement: The practice of replacing older, less efficient equipment or systems before the end of their expected operational life with newer, more energy-efficient models.

ECAM (Energy Charting and Metrics): A tool used to visualize and analyze building performance data.

ECM (energy conservation measure): An intervention intended to reduce a facility's energy consumption.

Econometrics: The use of statistics to study economic matters.

EEM (energy efficiency measure): Any machine, software, system, practice, or retrofit that leads to a general reduction in energy usage, without significantly impacting level-of-service can be called an energy efficient measure.

Effective useful life (EUL): The estimated period during which an energy-related product, system, or piece of equipment is expected to operate at its intended level of functionality and efficiency before it needs replacement.

Efficiency: The ratio of useful output obtained to the energy input consumed. A system or process is considered more efficient if it delivers more output for the same or less energy input. It is often expressed as a percentage, indicating how well energy is converted to its intended purpose without wastage.

Efficiency as a Service (EAAS): Efficiency as a service is a pay-for-performance, off-balance sheet financing solution that allows customers to implement energy and water efficiency projects with no upfront capital expenditure. The provider pays for project development, construction, and maintenance costs.

EM&V: Evaluation, measurement, and verification provides accurate, transparent, and consistent metrics – based on good data – that assess the performance and implementation of energy efficiency projects, programs, and portfolios of programs.

EMCS (also "BAS" and "EMS"): Energy management control system.

Energy: The capacity to do work or the ability to cause change. It can be stored (potential energy) or active (kinetic energy). Energy sources can be renewable (like solar or wind) or non-renewable (like coal or natural gas).

EEP: Energy efficiency project.

Energy intensity: It is measured by the quantity of energy required per unit output or activity, so that using less energy to produce a product reduces the intensity.

ENERGY STAR Portfolio Manager: An online tool created by the U.S. Environmental Protection Agency to benchmark and measure the energy and water usage of commercial buildings. It allows building owners and managers to input utility data and other relevant building information to track and assess energy and water consumption over time, compare the building's performance with similar buildings nationwide, generate reports and graphs to visualize and communicate performance, and earn ENERGY STAR certification for top-performing buildings.

ESCO: Energy service company.

ESPC (energy savings performance contract). A contractual arrangement between an ESCO and a customer, typically a building owner or a facility manager, in which the ESCO provides energy management services and solutions in exchange for a fee. Often, the ESCO implements one or more energy management measures, such as lighting upgrades or HVAC system improvements, and the customer "pays back" the ESCO a portion of the avoided costs over a specified period of time.

ESM: Energy savings measure.

Evaluation: The systematic process of assessing and determining the performance, effectiveness, and impact of energy-related initiatives, programs, or systems.

Evaluator: A professional or entity responsible for conducting assessments of energy-related initiatives, programs, or systems.

EVO (Efficiency Valuation Organization). A nonprofit organization whose flagship product is IPMVP, the International Performance Measurement and Verification Protocol.

Ex ante: A Latin term meaning "before the event." It refers to predictions, estimates, or evaluations made before the implementation of a particular energy project or intervention. Ex ante analyses provide forecasts of expected energy impacts, costs, or other outcomes based on models, historical data, or other relevant information, before actual measurements are available.

Ex post: A Latin term meaning "after the event." It refers to assessments or evaluations conducted after the implementation of a particular energy project or intervention. Ex post analyses measure actual outcomes and compare them to the ex ante predictions or other benchmarks to understand the effectiveness and impact of the initiative.

FEMP: Federal Energy Management Program.

Flexibility: An approach to energy management intended to reduce energy consumption during peak demand periods, often through demand-response programs. For example, a facility might reduce energy consumption during periods of high electricity demand by adjusting equipment schedules or temporarily shutting down non-essential equipment.

Flex markets: Energy markets that manage the variability and intermittency of renewable energy sources by enabling grid operators and energy market participants to buy and sell flexible capacity and demand responses in real time.

Forecast: "Forecast" and "predict" are two terms that are often used interchangeably, but they have slightly different meanings. A forecast is a prediction of future conditions or events, based on a combination of past data and current trends. Forecasts are often used in business and economics to predict things like sales figures, stock prices, and market trends. On the other hand, a prediction is an educated guess or estimate about what might happen in the future. Predictions can be based on past data and trends, but they can also be based on intuition, expert knowledge, or other subjective factors.

Guaranteed savings: A type of energy savings performance contract with a financial risk structure where the host facility arranges the financing. Project "savings" are "guaranteed" by the ESCO to cover the financial service payments.

Guideline: General information on approaches, methods, and procedures. A guideline is meant to inform, not dictate.

Impact (see "Results"): The assessed quantity of energy and/or financial units assessed through the M&V process. Impact refers to the effect or influence that an intervention has. The impact of an intervention can be positive, negative, or neutral.

Independent variable: A variable presumed to exert influence on a dependent variable.

Industry standard practice (ISP)[2] A term used to describe a technology or measure that is the typical equipment or commonly used practice.

Inference (vs. prediction)[3] Inference refers to the process of drawing conclusions or making deductions based on available evidence or data. It involves using observed data to make educated guesses about underlying patterns, relationships, or characteristics.

Interactive effects: The influence of an ECM on the energy use of other equipment in a facility beyond the measurement boundary.

Intervention: An intervention is an action or set of actions taken to bring about a desired change. This change can be at the individual, group, or societal level. Interventions can be targeted at a variety of issues, such as health, education, social welfare, and the environment.

IPMVP: International Performance Measurement and Verification Protocol.

ISO: International Organization for Standardization.

Lifecycle: The stages of energy asset or system's "life" from design to teardown.

---

[2]https://www.doe2.com/Download/CPUC/ExAnteProcess/ISPGuideBookv12_A_livingfinal.pdf:

[3]https://www.datascienceblog.net/tags/bayesian/:

Lifecycle cost: The total cost of owning, operating, maintaining, and disposing of an energy asset or system over its entire lifespan. Includes initial capital costs, operational and maintenance expenses, energy costs, potential replacement or upgrade costs, and any end-of-life disposal costs or salvage values.

NIST (National Institute of Standards and Technology): A U.S. federal agency that develops and promotes measurement standards, including those related to energy management, renewable energy, and grid modernization.

NIST's Building Life Cycle Cost Programs are software tools developed to provide an economic analysis of proposed energy and water management projects. These programs help users evaluate the cost-effectiveness of building-related projects with respect to their lifecycle costs.

Load shape: Refers to the pattern of variation in the demand for electricity or other forms of energy over a specific period.

Market transformation: Strategic intervention and efforts aimed at causing lasting changes in the structure or dynamics of a market, such as to promote the widespread adoption of energy-efficient practices and technologies.

M&V (measurement and verification): A process that includes a wide range of measurement, verification, and analysis techniques for quantifying the results of resource efficiency and resource management activities.

M&V 2.0: An advanced approach to M&V that leverages high-resolution data from advanced meters, smart building systems, and other Internet of Things (IoT) devices to provide more accurate and detailed analysis of energy consumption.

Measurement boundary: A conceptual line demarcating the space in which measurements of equipment and systems will occur.

Method: A specific way of carrying out a task or process. It is a concrete and detailed set of steps or procedures for achieving a particular goal. A method is typically more structured and prescriptive than an approach and often involves following a set of guidelines or procedures.

Metrologist: A professional specializing in metrology, the science of measurement.

Metrology: The science of measurement, applied to ensure the accuracy, precision, and reliability of instruments and methods used to quantify energy consumption, production, and performance.

Microgrid: A localized group of electricity sources and loads that can operate both connected to the main power grid or in an islanded mode, autonomously.

Model: Refers to a mathematical or computational representation that captures the relationship between various factors affecting energy consumption and the resulting energy usage.

Monitoring and targeting (M&T): The process of tracking energy consumption and setting specific goals or targets for energy management activities.

NegaWatt: A "negawatt" is a concept introduced by energy economist Amory Lovins in the 1980s. The term is a combination of "negative" and "watt," where "watt" is a unit of power or energy consumption. The concept of a negawatt refers to a unit of energy that is saved or conserved through energy efficiency measures, resulting in a reduction in energy consumption.

Net-to-gross (NTG): In a utility program, the net-to-gross (NTG) ratio indicates what portion of energy impacts can be directly attributed to the program itself. In essence, it is a measure of causality. It is calculated by dividing the net impact (the change in energy consumption attributable to the program) by the gross impact (the total observed change in energy consumption, regardless of cause). The resulting ratio ranges from 0 to 1, where 1 means that the program is directly responsible for all observed impacts, and 0 means that it is directly responsible for none.

Non-routine adjustment: Adjustments that must be made to account for changes to elements of the baseline that were not expected to change during the reporting period.

Non-routine event: Any event of sufficient magnitude (within the measurement boundary) not anticipated in the M&V plan and codified as a routine adjustment.

Non-energy benefits (NEBs): Other benefits that may result from the implementation of an energy management project, such as improved occupant comfort, indoor air quality, productivity gains, and reduced maintenance costs.

Normal: This word has many meanings. Whenever you see it in the context of M&V, make sure everyone agrees on the definition. A few specific cases follow.

Normalize: To adjust energy consumption data to account for variables or factors that might distort a direct comparison.

Normalized metered energy consumption (NMEC): NMEC is a methodology used to calculate energy savings achieved through the implementation of an energy efficiency measure or project. NMEC is intended to provide a clear mechanism to identify the savings achieved by measuring the real savings achieved based on the customer's reduction in energy use measured at the meter.[4]

Normalized savings: Calculated energy "savings" after adjusting for various external factors that might influence energy consumption. The use of the term "avoided energy/cost savings" or "normalized savings" is instantiated in the M&V guidance documents. We prefer to avoid using the term "savings." But in some cases, it is a term of art that an M&V professional must use to be understood.

Objective: See "Subjective."

Open source: Software or technologies whose source code is made available to the public, allowing anyone to view, modify, and distribute it. Open-source tools in energy management can include software for modeling energy consumption, tracking the impact of energy management activities, simulating building performance, and more.

Operational verification: The process of periodically verifying that an ECM is operating as intended through equipment inspection, review of operational logs, or other means.

Operations and maintenance (O&M): Routine practices and activities carried out so that equipment, systems, and facilities function efficiently and effectively – such as inspections, cleaning, calibration, repairs, and replacements. Proper O&M helps ensure that energy-consuming systems operate at their optimal efficiency, prolongs the lifespan of equipment, reduces downtime, and prevents excessive energy consumption or waste due to malfunctioning or poorly maintained equipment. A reduction in O&M expenses can also be a non-energy benefit of energy management activities.

Options: Standardized approaches or methodologies used to determine the impact of energy management activities.

Order of operations: The sequence of the analysis of ECM implementation. In many cases, the order of implementation can have a significant effect on the overall impact achieved, as well as the attribution of those impacts to individual ECMs.

---

[4]https://docs.cpuc.ca.gov/PublishedDocs/Efile/G000/M520/K881/520881077.PDF

Parameter: In energy modeling, a "parameter" refers to a variable or factor that can be adjusted or specified to represent certain characteristics of a system, process, or technology being analyzed. Parameters are essential elements in energy modeling as they allow the model to represent real-world conditions and behaviors. By modifying parameters, energy modelers can simulate different scenarios, assess the impact of various factors, and make predictions about energy consumption, efficiency, and other outcomes.

Parameter sweep: Systematic variation of input parameters to observe the resulting changes in outcomes, in order to understand the sensitivity of a model to changes in specific parameters.

Pay for performance: In the context of energy, "pay for performance" often refers to a type of energy efficiency program or policy where individuals, organizations, or businesses are rewarded financially based on the actual energy impacts they achieve. Utilities or government agencies may offer pay-for-performance programs to incentivize energy consumers to adopt energy-efficient measures. These measures could include upgrading lighting systems, improving insulation, installing efficient appliances, and more. Instead of providing upfront rebates or incentives, pay-for-performance programs provide financial rewards based on the actual energy impacts achieved over a specific period.

Payment (see also "Settlement"): Payment and settlement are related in that payment is often a part of the settlement process. Settlement refers to the process of resolving a financial transaction, which can include the payment of money or other compensation. For example, when a contract is signed for the sale of goods or services, payment is typically made as part of the settlement process to complete the transaction. After payment has been made, the transaction is considered settled.

Performance: In mechanical systems, performance refers to how well the system can function or operate according to its design specifications. This may include measures such as efficiency, power output, speed, accuracy, or reliability. The performance of a mechanical system can be affected by several factors, including the design and configuration of the system, the quality of the materials and components used, and the operating conditions under which the system is used.

Performance verification: Can be instantaneous, short-term, or long-term:

- **Instantaneous:** *Spot measurements* of the performance of individual components, such as HVAC systems, lighting systems, and other energy-consuming devices.
- **Short-term:** Periodically monitor energy consumption to track performance over a period of weeks or months – whatever is sufficient to evaluate a *performance curve*. This typically entails collecting data from energy meters or building automation systems and comparing energy consumption to expected or baseline values.
- **Long-term:** Monitor energy consumption and performance over a period of years, typically using an energy use index (EUI) or similar metric. The EUI is a measure of a facility's energy efficiency, calculated by dividing the total energy consumed by the facility by the facility's floor area: kWh per sq. ft. per year.

Physical model: Also known as a simulation or a causal, physics-based, or first principles model. A model that tries to simulate the underlying processes and interactions that give rise to the energy data, based on the idea of using mathematical or computational techniques to represent these processes.

Portfolio: A collection of energy efficiency programs sponsored by an organization such as a utility or government agency, employing various technologies across a range of contexts.

Power quality: Power quality refers to the characteristics of electrical power in terms of voltage, frequency, waveform distortion, and stability. A high-quality power supply should provide a stable and consistent voltage and frequency, free from disturbances and harmonic distortion. Poor power quality can result in issues such as voltage sags, surges, harmonics, flicker, and interruptions, which can adversely affect the performance and reliability of electrical equipment and systems.

Pre-post analysis: A pre-post analysis is a type of study that compares a group of subjects before and after an intervention. It is used to assess the effectiveness of a treatment or program by comparing the outcomes before and after the intervention.

Precision (vs. accuracy): How consistent the estimate is when the analysis is repeated multiple times.

Prediction: Prediction involves estimating or forecasting future outcomes or values based on existing data and patterns. It is about making an educated guess about what might happen in the future. Predictive models are built using historical data and patterns to make predictions about unknown or future instances.

Probability: The likelihood or chance of a particular outcome or event. It is a measure of uncertainty and is expressed as a number between 0 and 1, with 0 indicating that the outcome is impossible and 1 indicating that the outcome is certain. Probability is often used in statistical analysis to describe the likelihood of different events.

Productivity: The efficiency with which energy is converted into useful work or output in a system or process. Higher energy productivity indicates more value or output is derived from each unit of energy consumed.

Program: A series of related energy efficiency projects sponsored by an organization such as a utility or government agency.

Protocol: A standardized set of guidelines or procedures for the consistent and accurate measurement, reporting, and verification of the impact of energy management activities.

Note: While both guidelines and protocols provide guidance and instructions, guidelines offer broader recommendations that allow for flexibility and judgment, whereas protocols provide detailed, standardized procedures with less room for interpretation to ensure consistency and precision in executing specific tasks or processes.

Proxy variable: A substitute for an unobservable or difficult-to-measure variable that is believed to be correlated with the variable of interest.

Pulse counter: A device that counts and records the number of pulses or signals outputted by a meter or another measuring device. These pulses typically represent a fixed quantity of the parameter being measured, such as kilowatt-hours of electricity, liters of water, or cubic meters of gas. For example, an electricity meter might emit a pulse for every kilowatt-hour of electricity consumed. The pulse counter will then count these pulses, allowing for the tracking and recording of energy consumption over time.

Real time: The ability to measure energy usage or production at a particular moment in time and obtain immediate feedback on that measurement. Real-time measurements allow for monitoring of energy consumption or production in real time, which can be useful for identifying patterns, trends, or anomalies as they occur.

Real-time energy measurements are typically obtained through the use of sensors or meters that are capable of measuring energy usage or production at regular intervals, such as every second or every minute. The data collected from these sensors or meters can then be used to generate real-time energy consumption or production graphs, charts, or other visualizations that can be used to analyze and monitor energy usage or production in real time.

Realization rate: A metric that compares the actual observed impact of an energy management activity to the estimated or predicted impact. It is usually expressed as a percentage and helps gauge the effectiveness of ECMs.

Realization Rate = (Actual Impact/Estimated Impact) × 100%.

A realization rate of 100% indicates that the actual impact perfectly matched the estimated impact. Values greater than 100% mean actual impact exceeded estimates, while values less than 100% mean the ECM underperformed relative to expectations.

Regulations/regulatory: A domain of M&V that requires the M&V professional to understand, navigate, and comply with local, national, and international laws and regulations – whether they relate to energy efficiency, carbon emissions, environmental protections, or industry-specific requirements.

Remaining useful life (RUL): RUL refers to the estimated or projected amount of time that a specific system, component, or asset is expected to continue functioning efficiently and effectively before it reaches the end of its useful operational life. It is a critical consideration in assessing the long-term benefits of energy efficiency interventions.

Resource savings: Measurements of resource "savings" include electric energy (kWh) and capacity (kW) savings, natural gas savings (therms), and other resources (oil, propane, water, and maintenance), where applicable.

Regression: An analysis of the relationship between independent and dependent variables.

Reporting period: The period following the implementation of an ECM during which metering occurs.

Residual (analysis): A method used to evaluate the goodness of fit of a statistical model by examining the differences between the observed values and the predicted values (residuals). These residuals represent the deviation between the actual data points and the model's estimated values, and residual analysis involves assessing the distribution, pattern, and magnitude of these residuals to determine whether the model adequately captures the underlying data patterns.

Results: The assessed impacts generated through planned energy management activities.

Retrocommissioning: Often abbreviated as RCx. The process of reviewing and improving the operation and maintenance procedures of an existing building's systems and equipment to ensure they function optimally and efficiently. Unlike standard commissioning, which occurs during a building's initial setup or after a significant renovation, retrocommissioning focuses on existing buildings that have been in operation, often for years.

Retrofit isolation: An M&V approach that measures the energy use of the individual systems and equipment affected by a retrofit.

Rigor: A concept used mainly in evaluation study design (see CAL EMV protocol)[5].

Risk: The potential for loss or harm associated with a particular decision or course of action. It is a measure of the likelihood and magnitude of a negative outcome. Risk management is the process of identifying potential risks and developing strategies to mitigate or minimize them. In finance, risk is often measured using standard deviation or other statistical measures of variability.

Risk structure: The distribution of risk among project stakeholders.

Rounding: To adjust a number to a specified level of detail or decimal place.

Routine adjustment: An adjustment that reflects expected changes in independent variables – changes in weather from season to season, for instance. They occur often during the reporting period.

Savings: An imprecise concept often used to describe the value generated through energy management. It is potentially the most misused and dangerous word in energy management and M&V.

Settlement: See "payment."

Significant parameter: A variable or factor that has a notable or measurable impact on the outcome or performance of a system or model.

---

[5]https://www.cpuc.ca.gov/-/media/cpuc-website/files/legacyfiles/c/5212-caenergyefficiencyevaluationprotocols.doc

Simulation: A physical model becomes a simulation when it is used to replicate real-world behaviors, processes, or systems through experimentation or observation. The key distinction between a physical model and a simulation lies in the dynamic and interactive nature of the simulation, where the model is manipulated or operated to observe how it responds under different conditions.

Site-specific: An M&V plan tailored specifically to the unique characteristics, conditions, and requirements of a particular facility or system.

Smart meter: See "AMI."

Soft skills: Interpersonal and other non-technical skills, such as communication, problem-solving, and leadership, that enable individuals to work effectively with others, communicate their ideas clearly, and navigate the organizational and relational aspects of their roles.

Spherical cow: A humorous metaphor that highlights the tendency of some scientists and engineers to oversimplify complex problems to make them more tractable or easier to solve, often to the point of making their models practically useless for real-world applications.

Standard: A standard is a set of established norms, rules, or criteria that define how something should be designed, produced, operated, or executed. Standards provide a common framework for quality, safety, performance, and compatibility across different products, processes, or systems. Standards are developed through consensus among experts, industry stakeholders, and relevant organizations. They serve as benchmarks for best practices, allowing different entities to adhere to uniform guidelines.

Standard deviation: A statistical measure of how dispersed the data in a dataset is relative to the mean.

Standards of service: Predefined levels of service quality, reliability, and performance that energy services or utilities commit to providing to their customers.

Static factor: An element of the baseline not expected to change during the reporting period.

Statistical model: A model that attempts to identify correlations and other patterns in actual measured data.

Statistics: The collection, analysis, interpretation, presentation, and organization of data. It involves using mathematical methods to describe and analyze data and make predictions based on that data.

Super ESCO: According to the International Energy Agency (IEA), a governmental entity "created to serve the public sector, develop the capacity of private ESCOs, and facilitate project financing."

Targets: Specific goals or benchmarks set to achieve desired levels of energy consumption, efficiency, or emission reductions within a defined timeframe. *Targeting* is the process of setting, tracking, and striving to meet these energy-related goals.

Technical: The domain of M&V that involves the practical application of engineering principles.

Technical resource manual (TRM): A manual that includes a standardized set of methodologies, assumptions, and guidelines to estimate the impact of specific energy management activities. Utilities, program administrators, and regulatory bodies commonly use TRMs to ensure that energy management impact claims from various programs and initiatives are consistent and based on established methodologies.

Time series (data): A sequence of numerical values or observations collected at regular intervals over a period of time.

Timing of intervention: The strategic decision of when in the lifecycle of a facility to implement a particular ECM.

Uncertainty: The lack of knowledge or information about a particular outcome or event. It is the degree to which an outcome or event is unknown or unpredictable. Uncertainty is often measured using probability distributions, which assign probabilities to different possible outcomes.

Uniform methods project: A project managed by the National Renewable Energy Laboratory (NREL) for the U.S Department of Energy that aims to bring uniformity to the way states, utilities, program administrators, and evaluators calculate the impacts of energy management activities.

Universal translator: The UT is software designed for the management and analysis of data from loggers and trend data from building management systems (utonline.org).

Utility program: An energy efficiency program sponsored by a utility or government.

Valuation: The process of assigning a financial value to the impacts of energy management activities.

Value of information: The value of information refers to the benefit or advantage gained from obtaining new or additional information. In various contexts, such as decision-making, problem-solving, research, and business strategies, information can be considered valuable if it contributes to reducing uncertainty, improving the quality of decisions, increasing the likelihood of success, or enhancing understanding.

Verification (general): Confirming the operation and performance of equipment or processes in an energy efficiency project. It is also known as confirming the "ability to perform."

Whole facility approach: An M&V approach that measures the total energy use of a building or facility.

Index

About the Author

Steve Kromer P.E. is an energy engineer and principal at SKEE in Berkeley, California. With over three decades of experience in energy efficiency, Steve has been a leader in the development and deployment of measurement and verification around the world. He is a Certified Measurement and Verification Professional (CMVP) and a registered professional engineer in the State of California.